ISSUES IN ENVIRONMENTAL SCIENCE
AND TECHNOLOGY

EDITORS: R. E. HESTER AND R. M. HARRISON

3
Waste Treatment and Disposal

ISBN 0-85404-210-5
ISSN 1350-7583

A catalogue record for this book is available from the British Library

© The Royal Society of Chemistry 1995

All rights reserved
No part of this book may be reproduced or transmitted in any form
or by any means—graphic, electronic, including photocopying,
recording, taping or information storage and retrieval systems—
without written permission from The Royal Society of Chemistry

Published by The Royal Society of Chemistry, Thomas Graham House,
Science Park, Cambridge CB4 4WF

Typeset by Vision Typesetting, Manchester
Printed and bound in Great Britain by Bath Press, Bath

Preface

All of us are involved in generating waste; its treatment and disposal are essential to the maintenance of our society. In the previous issue of this series we reviewed waste incineration in considerable detail in relation to its environmental impact. We now turn to other options for waste management.

The first article, by the Director and Chief Inspector of HM Inspectorate of Pollution in the UK, David Slater, outlines the principles and practice of Integrated Pollution Control as established by the UK Environmental Protection Act 1990. This provides a discussion of BATNEEC and BPEO and includes sections on cleaner technology and waste minimization. Following this Peter Try and Graham Price of the Wessex Water Company examine the environmental impact of sewage and industrial effluent treatment, giving attention to discharges to inland and coastal waters, atmospheric emissions, and land disposal.

Landfill is an important waste management option which is reviewed by Ken Westlake of the University of Loughborough. His article examines current best practice in the design, engineering, operation, and control of landfill sites and identifies the effects of changing regulatory policy. Following this is an article on emissions to the atmosphere by Gev Eduljee of Environmental Resources Management. This puts the products of incineration into a broader context by considering the wide range of other waste management processes which result in releases to atmosphere, such as dusts, odours, fumes from industrial processes, pathogens, and microbial emissions from municipal wastes, *etc.* The regulatory framework is described and localized impacts and their mitigation are examined.

In the final three articles, recycling, nuclear fuel wastes, and the economics of waste management are reviewed. Steven Ogilvie and Jan Gascoigne of the National Environmental Technology Centre provide an insight into the general and technical issues facing the materials recycling industry. K.W. Dormuth, P. A. Gillespie, and S.H. Whittaker of Atomic Energy Canada Ltd. describe the qualitative and quantitative considerations which determine the suitability of deep underground storage for radioactive fuel waste produced by nuclear power reactors. David Pearce and Inger Brisson of University College London outline the principles of cost–benefit analysis which lead to a consistent economic theory of waste management, illustrating this with data from a number of case studies which demonstrate the utility of economic instruments in this area.

Preface

In conjunction with Issue No. 2 of this series, this present volume provides a thorough and detailed review of waste management options, incorporating considerations of environmental impacts and the main technical, regulatory, and economic factors which influence decision making. We are confident that this will be of value to scientists, technologists, legislators, consultants, and industrialists with waste disposal problems.

Ronald E. Hester
Roy M. Harrison

Contents

Integrated Pollution Control and Waste Minimization   1
D. Slater

1 An Introduction to Integrated Pollution Control   1
2 Process Regulation   3
3 Monitoring   4
4 BATNEEC   5
5 Published Guidance   8
6 Best Practicable Environmental Option (BPEO)   10
7 Cleaner Technology   11
8 Waste Minimization   14
9 Conclusion   15

Sewage and Industrial Effluents   17
P.M. Try and G.J. Price

1 Introduction   17
2 Regulatory Framework   20
3 Sewage Treatment   22
4 Industrial Effluents   32
5 Future Developments   35
6 Conclusions   40
   Acknowledgements   41

Landfill   43
K. Westlake

1 Introduction   43
2 Principles of Landfill Practice   45
3 Landfill Processes   50
4 Landfill Design, Engineering, and Control   56
5 Landfill Gas   59
6 Site Closure and Aftercare   62
7 Costs, Financial Provision, and Liabilities   64
8 Landfill—The Future?   65
9 Conclusion   67

Contents

Emissions to the Atmosphere   69
G.H. Eduljee

1 Introduction   69
2 Regulatory Framework in the UK   71
3 Characterization of Emissions   74
4 National and Global Impacts   78
5 Localized Impacts   82
6 Mitigation of Impacts   88

Recycling Waste Materials—Opportunities and Barriers   91
J.L. Gascoigne and S.M. Ogilvie

1 Introduction   91
2 The UK Waste Context   93
3 Recycling   96
4 General Issues Affecting Recycling   97
5 Technical Issues Affecting Recycling of Specific Materials   102
6 Materials Case Study: Options for Scrap Tyre Recycling   103
7 Conclusions   113

Disposal of Nuclear Fuel Waste   115
K.W. Dormuth, P.A. Gillespie, and S.H. Whitaker

1 Introduction   115
2 The Canadian Disposal Concept   118
3 Qualitative Discussions of Long-term Performance   121
4 Quantitative Case Study of Long-term Performance   124
5 Implementation of Disposal   128
6 Conclusions   130
    Acknowledgement   130

The Economics of Waste Management   131
D.W. Pearce and I. Brisson

1 The Waste Disposal Problem   131
2 The Economic Approach to Waste Management   131
3 Integrated Waste Management: An Economic Theory Approach   133
4 Measuring Waste Disposal Externalities   138
5 Economic Instruments for Integrated Waste Management   146
6 Conclusions   152

Subject Index   153

Editors

Ronald E. Hester, BSc, DSc(London), PhD(Cornell), FRSC, CChem

Ronald E. Hester is Professor of Chemistry in the University of York. He was for short periods a research fellow in Cambridge and an assistant professor at Cornell before being appointed to a lectureship in chemistry in York in 1965. He has been a full professor in York since 1983. His more than 250 publications are mainly in the area of vibrational spectroscopy, latterly focusing on time-resolved studies of photoreaction intermediates and on biomolecular systems in solution. He is active in environmental chemistry and is a founder member and former chairman of the Environment Group of The Royal Society of Chemistry and editor of 'Industry and the Environment in Perspective' (RSC, 1983) and 'Understanding Our Environment' (RSC, 1986). As a member of the Council of the UK Science and Engineering Research Council and several of its sub-committees, panels, and boards, he has been heavily involved in national science policy and administration. He was, from 1991–93, a member of the UK Department of the Environment Advisory Committee on Hazardous Substances and is currently a member of the Publications and Information Board of The Royal Society of Chemistry.

Roy M. Harrison, BSc, PhD, DSc(Birmingham), FRSC, CChem, FRMetS, FRSH

Roy M. Harrison is Queen Elizabeth II Birmingham Centenary Professor of Environmental Health in the University of Birmingham. He was previously Lecturer in Environmental Sciences at the University of Lancaster and Reader and Director of the Institute of Aerosol Science at the University of Essex. His more than 200 publications are mainly in the field of environmental chemistry, although his current work includes studies of human health impacts of atmospheric pollutants as well as research into the chemistry of pollution phenomena. He is a former member and past Chairman of the Environment Group of The Royal Society of Chemistry for whom he has edited 'Pollution: Causes, Effects and Control', (RSC, 1983; Second Edition, 1990) and 'Understanding our Environment: An Introduction to Environmental Chemistry and Pollution' (RSC, Second Edition, 1992). He has a close interest in scientific and policy aspects of air pollution, currently being Chairman of the Department of Environment Quality of Urban Air Review Group as well as a member of the DoE Expert Panel on Air Quality Standards and Photochemical Oxidants Review Group and the Department of Health Committee on the Medical Effects of Air Pollutants.

Contributors

I. Brisson, *Centre for Social and Economic Research on the Global Environment (CSERGE), University College London, Gower Street, London WC1E 6BT, UK*

K.W. Dormuth, *Atomic Energy of Canada Limited, Whiteshell Laboratories, Pinawa, Manitoba, Canada R0E 1L0*

G.H. Eduljee, *Environmental Resources Management, Eaton House, Wallbrook Court, North Hinksey Lane, Oxford OX2 0QS, UK*

J.L. Gascoigne, *AEA Technology, National Environmental Technology Centre, Culham, Abingdon, Oxfordshire OX14 3DB, UK*

P.A. Gillespie, *Atomic Energy of Canada Limited, Whiteshell Laboratories, Pinawa, Manitoba, Canada R0E 1L0*

S.M. Ogilvie, *AEA Technology, National Environmental Technology Centre, Culham, Abingdon, Oxfordshire OX14 3DB, UK*

D.W. Pearce, *Centre for Social and Economic Research on the Global Environment (CSERGE), University College London, Gower Street, London WC1E 6BT, UK*

G.J. Price, *Wessex Water Services Limited, Wessex House, Passage Street, Bristol BS2 0JQ, UK*

D. Slater, *Her Majesty's Inspectorate of Pollution, Room A538, Romney House, 43 Marsham Street, London SW1P 3PY, UK*

P.M. Try, *Wessex Water Services Limited, Wessex House, Passage Street, Bristol BS2 0JQ, UK*

K. Westlake, *Centre for Hazard and Risk Management, Loughborough University of Technology, Loughborough, Leicestershire LE11 3TU, UK*

S.H. Whitaker, *Atomic Energy of Canada Limited, Whiteshell Laboratories, Pinawa, Manitoba, Canada R0E 1L0*

Integrated Pollution Control and Waste Minimization

D. SLATER

1 An Introduction to Integrated Pollution Control

The introduction of the system of Integrated Pollution Control (IPC), proposed by the Royal Commission on Environmental Pollution (RCEP) in their Fifth Report[1] and embodied in the Environmental Protection Act (EPA 90) 1990, marked an important milestone in the development of the legislative philosophy and framework in the UK. The Act is of major importance since it largely established Britain's strategy for pollution control and waste management for the foreseeable future. The Act itself is divided into nine main parts. However, only Part I is relevant to the theme of this article.

Before the introduction of IPC under the Act in April 1991, emissions from major polluters to the three environmental media of air, water, and land were subject to individual and distinct control regimes. IPC provides a mechanism and a legal basis for looking at the impact which a process as a whole has on the environment as a whole. IPC takes a holistic approach, ensuring that substances which are unavoidably released to the environment are released to the medium to which they will cause the least damage. It embodies the precautionary principle: prevention is better than cure. As the saying goes: prevent, minimize, and render harmless. Also, the regulatory process, from application, through authorization, to regular returns of monitoring releases to the environment, and, where appropriate, to the enforcement action by Her Majesty's Inspectorate of Pollution (HMIP), is open to public scrutiny and comment.

Since the Act was first enacted in 1990, twelve Regulations have been made to specify and, at times, to vary the original requirements which it established in Part I. Typical, and perhaps the most important of these Regulations, are the Environmental Protection (Prescribed Processes and Substances) Regulations 1991 (Statutory Instrument SI/1991/472), and it is useful to examine these as an example of the purpose of the Regulations. As described above, Part I of the Act makes provision for integrated pollution control (IPC). It also makes provision for the control of air pollution by local authorities.

Regulations SI/1991/472 also provide a framework for the implementation of a

---

[1] 'Fifth Report by the Royal Commission on Environmental Pollution', HMSO, London, 1976, Cmnd 6371.

substantial number of EC Directives relating to air pollution from industrial plants. It should be noted that since SI/1991/472 came into force in April 1991 in England and Wales, and in 1992 in Scotland, it has been modified six times. These amended Regulations, in general, extend (or at least redefine) the prescribed processes covered by the original Regulations and the dates for their authorization.

IPC applies to all processes in England, Wales, and Scotland falling within any descriptions of processes prescribed for the purpose by the Secretary of State for the Environment. The Act provides that no prescribed process may be operated without an authorization from HMIP after the date specified in the regulations for that description of process.

In setting the conditions within an authorization for a prescribed process, Section 7 of the Environmental Protection Act 1990 places HMIP under a duty to ensure that certain objectives are met. The conditions should ensure:

(i) that the best available techniques not entailing excessive cost (BATNEEC) are used to prevent or, if that is not practicable, to minimize the release of prescribed substances into the medium for which they are prescribed, and to render harmless both any prescribed substances which are released and any other substances which might cause harm if released into any environmental medium;

(ii) that releases do not cause or contribute to the breach of any direction given by the Secretary of State to implement European Community or international obligations relating to environmental protection, or any statutory environmental quality standards or objectives, or other statutory limits or requirements; and

(iii) that when a process is likely to involve releases into more than one medium (which will probably be the case in many processes prescribed for IPC), the best practicable environmental option (BPEO) is achieved (*i.e.* the releases from the process are controlled through the use of BATNEEC so as to have the least effect on the environment as a whole).

HMIP is also charged with delivering the National Plan for reduction of SO_2 and NO_x emissions by means of the authorizations which it grants under Part I of the Environmental Protection Act 1990. This translates a blanket concept, which takes no account of the pollution potential of an individual plant, into a site-specific allocation, the use of which can be accounted for by the operator, audited by HMIP, and enforced against if necessary.

The processes covered by the National Plan can broadly be grouped under three headings: first, the electricity supply industry; second, the petrochemical industry, comprising the refineries; and third, other industry, which picks up the power-generating combustion processes of 50 MW input or more.

There is often confusion about whether the National Plan takes precedence over BATNEEC and BPEO, or *vice versa*? The answer is simple. All objectives have to be achieved, so effectively it is the most stringent which will prevail. If BATNEEC standards are the tighter, then BATNEEC is pre-eminent. If BATNEEC would allow greater releases than the National Plan allocation, then National Plan prevails.

2 Process Regulation

The Environmental Protection Act 1990 requires an 'Authorization' to be issued by the 'Enforcing Authority' which, in the case of IPC, is Her Majesty's Inspectorate of Pollution. 'Authorization' means an authorization for a process (whether on premises or by means of mobile plant) granted under Section 6 of EPA 90.

The main goal, therefore, of an authorization is to specify the limits and conditions which are important to achieving the objectives of IPC in a particular circumstance. These are in the main likely to comprise limits and conditions on feed materials, operating parameters, and release levels. In turn, the detail of these, particularly of the latter, such as the period over which they apply, will need to take into account what constitutes BATNEEC, environmental impact, *e.g.* concentration or load-dependent process characteristics, *e.g.* cyclic variations, fluctuations, and practical considerations.

The operator must provide a strong, detailed justification of his process. Particularly where it is new or modifications are proposed, then all the options must be explored and justified. He must list the substances which might cause harm, that are used in or result from the process. He must identify the techniques used to prevent, minimize, or render harmless such substances. He must assess the environmental consequences of any proposed releases. He must detail the proposed monitoring of the releases. This comprehensive environmental assessment then becomes the corner stone of the assessment of the application by HMIP and available to all on the public register. HMIP can also be judged, by the public, on the way it responds. The public should then have confidence in the system.

The authorization document follows a standardized format, produced within HMIP. The conditions in the authorization have been set to ensure a consistency from one authorization to another. The standard authorization format is set out with standard assumptions. These are that the Applicant should be an expert in his own process, and that the Inspector is the expert on the requirements of the regulations and on assessment. The Applicant will have a much better understanding of his operations, requirements, process control, and variations in conditions than will the Inspector. It is for the Applicant to detail, where relevant to his process, the practicable conditions and assess the environmental implications. It is for the Inspector to assess the validity and acceptability of the proposals in respect of statutory environmental requirements. The application itself forms part of the authorization.

Releases to the various media are dealt with in a consistent manner in separate parts of the authorization. A plan should be included in the application which identifies the position of the various release points. The Inspector must consult with the National Rivers Authority (NRA) with respect to discharges to 'Controlled Waters' from the process, and include as a minimum requirement any conditions which the NRA insist shall be included. HMIP may impose conditions which are more onerous than those required by the NRA (*e.g.* continuous monitoring instead of spot sampling). Inspectors must check that the criteria used by NRA are appropriate. Section 28(3)b of the EPA requires that the inclusion of NRA's requirements are conditions of the authorization.

The Waste Regulatory Authority will be responsible for final disposal of waste to land (including on-site final disposal) in accordance with their licensing system. HMIP is responsible for control of any releases to land not subject to control by the Waste Regulatory Authority.

An authorization includes an Improvement Programme. For a new process the improvement programme may contain requirements for additional equipment/controls, *etc*. The Applicant must demonstrate that BATNEEC applies during the interim period (*e.g.* cost of lost production *versus* environmental effects if there is a delay or an extended period in delivering equipment). The regulation of releases will be enforced by regular site inspections, checking against authorization conditions. The operator will be required to provide information to HMIP as specified by the authorization conditions. Whenever breaches of authorization conditions are identified, due consideration will be given to enforcement actions that could lead to prosecution in the Courts.

3 Monitoring

Unfortunately, the word 'monitoring' is open to wide interpretation. In its general sense it includes all HMIP's regulatory functions, including physical inspection of processes and plants, the environment around the sites, and the paperwork, *e.g.* operators' returns, *etc*. However, in the context of authorizations we are referring to 'compliance monitoring' involving the measurement and recording of those aspects of a process which are subject to limitation or condition and 'environmental monitoring' when this is required as a condition. Although narrower than the general interpretation, 'compliance monitoring' can still encompass a wide variety of measurements.

When determining compliance monitoring requirements under IPC, Inspectors consider the following points.

(i) *What information is required to provide confidence that the process is being operated within authorized limits and conditions?* These requirements must be capable of providing data for the period over which the limit is expressed. Thus, if the limit is 'instantaneous', then a spot sample (or measurement) will be specified. If the limit is an average concentration expressed over 24 hours, then a 24-hour sample (or measurement) will be specified. If compliance with a load limit is to be monitored, then not only must the sample be taken over the period of the limit but a measure of flow will also be required.

(ii) *How might the information best be obtained?* Could more reliable data be obtained by deduction based on a knowledge of impurity levels in feedstocks and their behaviour during processing, *e.g.* mercury in caustic? Or is measurement the best option? If so, are on-line continuous monitoring systems available—do they constitute BATNEEC for that process? If reliance has to be placed on laboratory analysis of samples, how are they to be taken—by continuous sampling on a flow or time proportional basis, and over what time, or if by periodic sampling at what frequency and duration? Might surrogate measurements be more practicable?

For example, for dioxin emissions from incinerators, continuous monitoring is not available; periodic testing should be called for but is too expensive to employ more than perhaps once per month. Measurements of temperature, flow rates, feed composition, *etc.*, would provide valuable information on whether an incinerator of acceptable design has been complying with dioxin emission limits.

EPA 90, Part 1, places no statutory requirement on either operator or regulator to carry out monitoring beyond providing Inspectors with rights of entry and powers to take away samples, *etc.* However, monitoring is fundamental to providing information to the Inspectorate and the general public:

(i) on whether operators are within authorized limits and conditions; and
(ii) on the levels of releases actually being made to the environment.

In its purest form an authorization need not refer to monitoring at all. The essential point is to set limits and conditions which, if complied with, constitute BATNEEC and provide for satisfactory environmental protection.

The approach that has been adopted under IPC is to place requirements on operators to monitor their releases, to keep the results of their monitoring for inspection, and to report key results to the Inspectorate for inclusion on publicly accessible registers. Furthermore, operators are required to use the best available techniques for monitoring, which HMIP regards as on-line instrumentation linked to computer data storage systems wherever practical and not entailing excessive cost.

There are arguments for independent measurement of releases but, with the growing complexity and sophistication of monitoring techniques, there are advantages in requiring the operator to carry out specified measurements of releases and to record them with other details. The regulator must make sure that these are done conscientiously and by way of tamper-proof controls and recording systems. This is part of the inspection task. In addition, confirmation and reassurance that this system is working honestly and effectively is provided by HMIP commissioning check-measurements. The purpose of the latter is not to duplicate or augment the information supplied by the operator but rather to provide information, independently generated, against which the operator's data may be compared.

This approach to monitoring has been portrayed as self-regulation. It is not. Rather, it represents a system which is rigorous, cost-effective for HMIP and operators, requires the operator to take a close interest in the environmental impact of his operation, and will provide increased and better quality data to the public on releases.

4 BATNEEC

The Environmental Protection Act lays a specific duty upon the Chief Inspector

'to follow developments in technology and techniques for preventing or reducing pollution of the environment due to releases of substances from prescribed processes'.

To discharge this duty, the Department of Environment allocates a sizeable research budget to HMIP, which is used to conduct and commission the research necessary to underpin HMIP's regulatory role. An important area of this research is to find out about available techniques which are in use in the UK and abroad in the process in question, and to evaluate their relative merits. A former Secretary of State for the Environment saw IPC as a driving force for innovation, creating and satisfying demand for new technologies not only in terms of abatement equipment but in terms of the whole process. What IPC seeks to achieve is the right balance between cost to industry of better processes and abatement equipment, and cost to the environment in terms of the damage caused by releases.

As was described earlier in this Article, all IPC processes under Part I of the Environmental Protection Act 1990 are subject to control via the philosophy of BATNEEC, which has been a part of European Community legislation for some time, having been first introduced in 1984 in Directive 84/360/EEC on combating air pollution from industrial plant. This states that

'... an authorization may be issued only when the competent authority is satisfied (among other things) that all appropriate preventive measures against air pollution have been taken, including the application of best available technology, provided that the application of such measures does not entail excessive costs ...'.

For an overall definition of BATNEEC it is helpful to consider the words separately and together. Thus:

- **'Best'** must be taken to mean most effective preventing, minimizing, or rendering polluting releases harmless. There may be more than one set of techniques that achieves comparable effectiveness—that is, there may be more than one set of 'best' techniques. It implies that the technology's effectiveness has been demonstrated.
- **'Available'** should be taken to mean 'procurable by any operator of the class in question'; that is, the technology should be generally accessible (but not necessarily in use) from sources outside as well as within the UK. It does not imply a multiplicity of sources, but if there is a monopoly supplier, the technique will count as being 'available' provided that all operators can procure it. Furthermore, it includes a technique which has been developed (or proven) at a scale which allows its implementation in the relevant industrial context and with the necessary business confidence. Industry often believes that for a technology to be recommended as 'available', it would need to have been in commercial use for at least six to twelve months.
- **'Techniques'** is a term which embraces both the plant in which the process is carried on and how the process is operated. It should be taken to mean the components of which it is made up and the manner in which they are connected together to make the whole. It also includes matters such as numbers and qualifications of staff, working methods, training and supervision, and also the design, lay-out, and maintenance of buildings, and will affect the concept and design of the process.

The other North Sea States have also adopted (at least in general) the concept of BAT(NEEC) and are in the process of applying it to industrial sectors discharging hazardous substances by reviewing discharge consents. Consents specify BAT(NEEC) in terms of treatment efficiency rather than specifying the actual technology to be used, so as not to hamper innovation, and these are continuously revised to include the latest developments.

As examples of the application of BATNEEC, in Sweden when consents are issued for new or enlarging plants, they also require research and development to further improve treatment technology. The Netherlands and Germany have set two levels of technological requirements. Thus in the Netherlands, Best Technical Means (BTM) applies to all potential EC List I Substances (Black List) and Best Practical Means (BPM) for all List II Substances (Grey List). BTM, in this case, is defined as the most advanced treatment technology which is practically usable, whereas BPM refers to technologies which are affordable for 'averagely profitable companies'. In Germany, Best Available Technology (BAT), similar to BTM in the Netherlands, must be applied to effluents containing dangerous substances, whereas Generally Acceptable Technology (similar to BPM in the Netherlands) must be applied to effluents containing biodegradable substances.

- **'Not entailing excessive cost'** needs to be taken in two contexts, depending on whether it is applied to new processes or existing processes. For new processes the presumption is that the best available techniques will be used. Nevertheless, in all cases BAT can properly be modified by economic considerations where the costs of applying the best available techniques would be excessive in relation to the nature of the industry and to the environmental protection to be achieved.

In relation to existing processes, HMIP seeks from applicants and operators fully justified proposals for the timescale over which their plants will be upgraded to achieve the same standards as would be expected of new plant. This is why for new gas turbine power stations HMIP has not required the fitting of relatively very expensive selective catalytic reduction in most circumstances. Very good control of emissions of nitrogen oxides is achievable by combustor and burner design. It also fits the Best Practicable Environmental Option (BPEO) criteria referred to later.

Reasonableness comes into all of the discussions on BATNEEC, in that a situation is achieved where money is spent on abatement only up to the point where the resulting increase in control of pollution justifies the money spent.

So far as the ground rules are concerned, HMIP has been proceeding on the presumption that coal-burning baseload stations and orimulsion stations would have to be fitted with flue gas desulfurization (FGD) or equivalent, low NO_x burners or equivalent, and high efficiency particulate arrestment. Stations on load factors below baseload should be subject to BATNEEC evaluation which balances industry-wide financial considerations against station-specific environmental impacts.

BATNEEC judgements refer to what is achievable by way of pollutant release levels by the application of process/abatement options which are accepted as representing best practice in the relevant context. Ideally, Inspectors' judgements for a specific site should be based on an awareness of financial and economic implications for the affected industry(ies) of what is identified as 'Best Practice', which will enable HMIP to gauge how far and how fast the regulated community can undertake environmental expenditures.

For the choice of production techniques, the options are the alternative ways of producing a specified output either for intermediate use or final consumption. The focus of this analysis is on assessing the costs of each technique against the associated environmental effects, so the options may usefully be identified in terms of the ranked environmental performance of each technique. The analysis can then seek to determine either the total or the incremental net cost of each option.

The method of appraisal outlined here aims to identify for a typical firm the incremental costs (unit or total) associated with using different technologies, and thus different levels of environmental impact, in order to produce the same final output. The comparison could be made in terms of, say, the difference in net present value of annual output of 1000 units for ten years; or the difference in unit costs in a base year, assuming full accounting costs are covered over project life.

A complete cost–benefit analysis would aim to identify and compare the full social costs of two or more different states of the world, so that a decision could be based on the change in Net Social Benefit (*i.e.* the estimated money value of the material gains and losses to all actors) between the different options. Hitherto, it has not been possible for Inspectors' judgements to be based on a full cost–benefit analysis because no reliable, cardinal monetary valuations are available for the environmental outcomes; the ranking is ordinal only. Hence there is no explicit attempt made to equate marginal costs and marginal benefits of environmental improvements. The BATNEEC solution is defined solely in terms of the total or incremental resource cost of implementing that environmental outcome.

Cost effectiveness is the principal economic criterion for identifying the BATNEEC solution. This requires comparison of the relative costs to the operator of alternative environmental quality levels, as determined by the feasible set of production techniques.

5 Published Guidance

The Inspectorate recognizes that the requirements of the Act, its regulations, and the procedures to be followed under it are demanding for an applicant for an authorization. Comprehensive guidance has therefore been prepared. General guidance is provided by the publication 'Integrated Pollution Control—A Practical Guide'[2] which

   (i) outlines the origins of IPC;
  (ii) explains authorizations, and the application and consultation procedures;
 (iii) explains the meaning of BATNEEC and how it is promulgated through Chief Inspector's Guidance Notes;
 (iv) discusses topics such as authorization variation procedures, fees and

---

[2] 'Integrated Pollution Control: A Practical Guide', HMSO, 1993.

charging, the functions of HMIP enforcement, appeals, and the interface with other legislation; and

(v) lists other information such as the prescribed substances and HMIP locations.

In order that all his Inspectors apply the current BAT standards consistently, the Chief Inspector issues guidance to them in the form of published 'Chief Inspector's Guidance Notes' (CIGN). The notes are available to the public and, of course, to operators and developers of prescribed process, so that they have a clear understanding of what are the necessary standards of operation—including the substances released and the emission/release levels that may be stipulated in a specific authorization.

Further general guidance to inspectors on technical matters is published in the form of Technical Guidance Notes, covering such matters as dispersion calculations, chimney height determinations, *etc*. These, like the process guidance notes, are publicly available through Her Majesty's Stationery Office (HMSO). HMIP's published guidance is listed in its bibliography.[3]

The legislation does not contain numerical standards. If the best available techniques enable an operator to prevent a release, and he can employ these techniques without entailing excessive cost, then that is what he will be required to do. Any published guidance is fixed in time, but the yardstick for authorizing his process is dynamic: BAT can change from day to day. It will be up to industry to drive the standards. Industry, whether it be individual companies or their trade associations, must keep abreast of developments.

The CIGNs to Inspectors have no statutory force. They do, however, represent the view of HMIP on appropriate techniques for particular processes, and are therefore a material consideration to be taken into account in every case. HMIP must be prepared to give reasons for departing from the guidance in any particular case. The final decision on a particular application will be taken following consideration of the Applicant's case and of any representations from the public and the statutory consultees.

Similarly, an applicant must not feel constrained by the Guidance Notes: if he deems that a particular technique constitutes 'best available' in the context of his process, it is for him to put that proposition forward, and to justify it in the terms of the Act. He might choose another technique which will deliver the same environmental performance as those mentioned in the Guidance Notes; or he might choose a route which will deliver a better performance, and therefore must be regarded as BATNEEC.

Both HMIP and industry must be constantly re-evaluating the standard, the techniques, and the economics relevant to any process. It is this dynamic feature of IPC which will enable sustained and sustainable environmental improvements to be achieved. HMIP is committed to making regular revisions of the CIGNs. As already stated, HMIP conducts and commissions research to find out about available techniques. It is this research which underpins the CIGNs to Inspectors. HMIP reviews and evaluates the techniques available and in use, in the UK and abroad, for the process.

---

[3] HMIP Bibliography, HMIP, London, May 1994.

Because BATNEEC is a site-specific concept, CIGNs now provide information relating to achievable release levels for new processes applying the best combination of techniques to limit environmental impact in the context of the processes described. This is aimed at overcoming possible confusion about the expression of release limits in earlier CIGNs which were occasionally interpreted as uniform emission standards. This is, of course, contrary to the nature of BATNEEC.

HMIP are currently considering the next steps for the development of CIGNs which aim to incorporate general sectoral advice on economic and market factors which should be taken into account in the assessment of BATNEEC. The programme for the revision of CIGNs commenced with the Fuel and Power sector during 1994/95.

6 Best Practicable Environmental Option (BPEO)

The Act requires that BATNEEC is used to achieve the Best Practicable Environmental Option (BPEO). So what is BPEO? The concept of BPEO was first introduced by the UK's Royal Commission on Environmental Pollution which, in 1988, described BPEO as

> 'the outcome of a systematic consultative and decision-making procedure which emphasizes the protection and conservation of the environment across land, air, and water. The BPEO procedure establishes, for a given set of objectives, the option that provides the most benefit or least damage to the environment as a whole, at acceptable cost, in the long term as well as the short term.'[4]

To authorize an IPC process in accordance with the BPEO (and BATNEEC) objective, it is necessary to compare alternative options for operating a process, and consequent releases, to ensure that one environmental medium is not protected to the detriment of another, and that the impact on the environment as a whole is minimized.

Under IPC, the requirement to satisfy the BPEO objective for such a process has caused HMIP to re-examine the way in which it has traditionally assessed applications for authorization. For example, how does the Inspectorate judge whether or not hydrocarbons should be flared of recycled, or whether or not heavy metals should be disposed of by means of landfill rather than discharged to an estuary? The primary aim must be to identify the maximum concentration of contaminants which can be tolerated in any portion of the environment, considering in particular the most sensitive pollution receptors. For regulatory purposes, this concentration or level can be regarded as an indication of the Environmental Capacity, *i.e.* the ability of a particular portion of the environment to tolerate (an increase in concentration of) a specific contaminant. Furthermore, the proportion of the environmental capacity utilized by a release can be used to indicate the relative harm caused.

Many decisions on IPC applications have been made without the scientific

---

[4] 'Best Practicable Environmental Option. Twelfth Report by the Royal Commission on Environmental Pollution', HMSO, London, February 1988, Cmnd 310.

methodology which we all recognize as being so very important. EPA 90 is new legislation which takes a very bold approach and embodies some adventurous objectives. BPEO was, when the Act was passed, a slightly abstract concept which has needed a great deal of development. It is an evolutionary approach, and will bear fruit over many years if not decades. It is a piece of legislation which enables us to apply the very best science and the most up-to-date thinking to the dual challenge of maintaining a healthy business base in this country and of affording the right level of protection to the environment.

Development of an environmental assessment procedure for BPEO began in 1992–93, and was continued strongly during 1993–94, with particular emphasis on consultation. In July 1993 the approach was presented at a seminar attended by nearly 100 delegates from industry, government, and environmental organizations, and further discussions have been held with industry groups on other occasions. The procedure has been significantly revised and extended on the basis of comments received, and was circulated as a consultation document to several hundred interested organizations and individuals in 1994. It is intended that following this consultation exercise a Guidance Note on principles for environmental assessment will be produced in 1995.

7 Cleaner Technology

Cleaner technology is about minimizing the environmental impact of releases from processes. The philosophy behind it is the prevention of waste rather than the cure. Every aspect of a process needs to be optimized to minimize waste in any form.

The basic options to achieve this philosophy are relatively few and can be summarized in order of preference as:

(i) *Reduction at Source*—The most effective way to prevent a material from entering the environment is to stop using or making it.
(ii) *Product Changes*—A process should only be operated if the products can not be made in a cleaner way. Suitable alternative materials may perform the same function with less environmental consequences.
(iii) *Process Changes*—A process should be designed or changed in such a way that potentially polluting materials are not made or isolated, minimizing the possibility of a release.
(iv) *Re-use*—Re-use of a material is an alternative way of preventing release to the environment.
(v) *On-site Recycling*—Using a by-product of one process as a raw material for another disposes of it without an environmental impact.
(vi) *Off-site Recycling*—Sending a by-product of a process to be used elsewhere is similar to on-site recycling, but the pollution and cost of transport, handling, *etc.* makes this less desirable.

In the event of it not being possible to prevent a pollutant being formed, it must be treated or destroyed to render it harmless. This, of course, is not cleaner technology.

Only after achieving all the possible moves up this scale do you move on to looking at the sort of end-of-pipe methods which are suitable for dealing with those wastes which you have failed to eliminate. And here, the sort of alternatives to be considered might include:

- Incineration;
- Chemical transformation to a less harmful waste;
- Biological treatment;
- Transfer from one environmental medium to another where it might be less harmful;
- Dilution or dispersion.

Cleaner technology is achieved by good engineering design, good management practices, and innovative process design. The actual cleaner technology will depend upon the industry or process concerned. For example, water conservation and energy management may be the most significant considerations for a company manufacturing soft drinks, while to chemical manufacturers raw material controls and synthetic route could be more important. Cleaner technology is thus the most efficient process, ensuring maximum utilization of all raw materials and energy.

Because cleaner technology is about innovative ideas for processes and proper management of people and equipment, it is not necessarily expensive technology. Cleaner processes can be operated by any organization, no matter what the size or type of operation. For many processes the hardware for cleaner operation already exists. What is required is the ingenuity to assemble the appropriate building blocks.

The advantage of using cleaner technology is as much financial as environmental. By designing a process to minimize waste, product yields are usually higher. On some occasions the extra capital costs offset the advantage of increased process efficiency but, by the time disposal costs of waste is taken into account, clean processes are normally economically advantageous.

Organic solvents which give rise to Volatile Organic Compounds (VOCs) are widely used in the manufacture and application of industrial paint systems. Paint manufacturers are developing water-based paint systems which do not contain organic solvents. This is an example of substitution. At present these aqueous systems are generally more expensive to produce. However, being non-flammable they present fewer safety hazards. The operator is not required to take as many safety precautions or fit as much abatement equipment to the paint shop. The cost of application in these circumstances will be less.

The power generation industry is an example where an innovative solution to a difficult problem has given rise to a cleaner technology. Most fossil fuels contain significant quantities of sulfur which is converted to sulfur dioxide during the combustion process. Sulfur dioxide is one of the major gases that gives rise to acid rain. The current practice of removing sulfur dioxide from the flue gas—FGD—prevents air-borne pollution but leaves a solid waste product instead. This meets the current requirements to reduce sulfur dioxide releases but is not clean technology. A new process which removes the sulfur prior to combustion is a much cleaner technology with many potential operational advantages.

This process is called an Integrated Gasification and Combined Cycle (IGCC) system. Coal (or oil) is gasified using one of several proprietary processes. The technology for this stage is well established and has been used in the chemical industry for many years. The synthesis gas generated has the sulfur contaminant present as hydrogen sulfide. The gas volume prior to combustion is about one hundred times less than if the same fuel were burned conventionally. The size of the gas cleaning plant consequently is smaller and less expensive in relation to an FGD plant. The process for removal of hydrogen sulfide from the synthesis gas also is well-established, having been used in the petroleum industry for many years. Sulfur is removed as elemental sulfur which has a ready commercial market. Removal of sulfur as the element is not possible if it has been through the combustion process and been oxidized to sulfur dioxide. Power is generated by burning the clean synthesis gas in a gas turbine working in a combined cycle mode. This part of the overall process is also well established, with many Combined Cycle Gas Turbine stations running on natural gas. The IGCC method of power generation is not only more efficient at converting the fuel to power than a conventional steam-raising power station, but much lower releases of nitrogen oxides (another acid-rain-producing gas and strong irritant) are also possible. Thus, for a given unit of fuel, more power is produced and less of each of the major pollutants is released. This is an illustration of innovative clean technology. Several demonstration plants are being built or planned and this process is the one most likely to be used for power generation in the early part of the next century.

In the case of companies that have made applications for IPC authorization, a striking example of one where a 'dirty' process has been replaced by 'clean' technology is that of a petrochemical company in Southern England. The company required pure butenes for its down-stream processes but the butene supply was heavily contaminated with butadiene. They had a requirement for butadiene and so operated a wet extractive process using cuprous ammonium acetate to separate the butadiene from the butenes. This process typically released annually around 200 tonnes of ammonia and 140 tonnes of VOCs to air, and around 300 tonnes of ammonia and 6 tonnes of copper to the aqueous effluent treatment system. Involving HMIP at the design stage, they resolved to source their butadiene from elsewhere and replace the butene purification process with a new one having essentially zero emissions. Authorization has been given for a catalytic hydrogenation process which selectively reduces the butadiene to the required end-product butene. There are no emissions to air other than those estimated for fugitive releases from flanges, valves, pump glands, *etc.*, (7 tonnes annum^{-1} of VOCs) and they estimate that less than 1 kg year^{-1} of hydrocarbons is released to the water course.

Another example of a company changing the process completely to eliminate the release of a prescribed substance—in this case VOCs—is that of an American-owned manufacturer of fluorescent tubes. The company is currently preparing an application for authorization as a 'Mercury process'. The insides of fluorescent tubes are coated with a phosphor that has traditionally been applied as a solution or suspension in a xylene/butanol-based solvent. Following the drying/curing process, around 500 tonnes year^{-1} of solvent is lost to the atmosphere. Having come under near-simultaneous environmental pressures on both sides of the Atlantic, the laboratories of the parent company put around 30 man-years of effort into finding a 'cleaner' process. They have now developed a water-based carrier system for phosphor deposition which is to be installed in the UK factory. The only significant release expected is 1.25 tonnes year^{-1} of ammonia loss to the atmosphere. The capital cost of this change of technology is high so the company is replacing its four existing production units in a four-year rolling programme.

8 Waste Minimization and IPC

In its widest context, waste can be interpreted as almost any loss or discharge of any material to any medium, and with particular emphasis on specified ranges of substances for the three environmental media. HMIP uses that interpretation in IPC.

In the issues of the Chief Inspector's Guidance Notes covering the Chemical Industry, Inspectors also are advised to encourage applicants to carry out a formal process assessment and have in place a Waste Minimization programme, in advance of submitting their application. In addition to identifying, in a systematic way, those areas in their process where reductions in releases may be accomplished, and thereby providing the foundations for a programme for up-grading the plant, the procedure provides a mechanism whereby applicants can identify deficiencies in the data to be included in their application (which is a problem HMIP has encountered far too frequently). The introduction of the waste minimization concept into those Chief Inspector's Guidance Notes that were issued after the Institute of Chemical Engineers' 'Waste Minimization Guide'[5] was published in 1992, is no coincidence. The appearance of the Guide provided a mechanism (with defined methodology) by which HMIP could encourage operators of IPC processes to take the next step towards the basic principle of IPC—namely, 'Prevention rather than cure'.

Under IPC, strictly it is only releases of substrates prescribed in the Regulations which must be prevented or minimized; other releases need only be rendered harmless. However, the range of prescribed substances, particularly those prescribed for air, is large, so even if waste minimization is restricted to these areas, very significant improvements can be made—and, once the thinking has started, it is unusual for the programme to be restricted solely to prescribed substances.

HMIP, along with the BOC Foundation for the Environment, the NRA, and Yorkshire Water Services, also were sponsors of a project on Waste Minimization, co-ordinated by the Centre for Exploitation of Science and Technology. This project concentrated on water treatment for a number of companies in the Aire and Calder catchment area in Yorkshire. The project clearly demonstrated environmental and economic benefits. Most of the benefits were as a result of the reduction in the use of inputs such as water, energy, and raw materials.

The Aire and Calder project has proved to us all that to prevent, to minimize,

[5] 'Waste Minimization Guide', The Institution of Chemical Engineers, Rugby, 1992.

Integrated Pollution Control and Waste Minimization

and to render harmless our releases to the environment is a readily achievable objective. It is the basic objective of IPC, and seems to me to be a very good one to apply to any of our activities. The Aire and Calder project has already shown us that there are shortcomings in measurement of releases to the environment even at the most modern sites. And as the project has progressed it has become clear that those companies which have wholly embraced monitoring strategies are now in control and can therefore formulate programmes for management, containment, and minimization of releases to all three environmental media. So we have a way in to what must be a continuous cycle of measurement, analysis, control, and feedback.

The project has demonstrated pay-back at three levels. First order savings are those related to good housekeeping. Second order savings come from an analysis of product losses and de-bottlenecking. And third order savings will come from a change in ethos from end-of-pipe techniques to inherently 'clean' processes. The conclusion is that pollution prevention pays.

The Centre for the Exploitation of Science and Technology has been commissioned to extend the concept by identifying three suitable areas for further studies. The new studies will be concentrated on industries and regions of the country that have not been subject to this approach with particular reference to those that will soon become subject to IPC regulation.

9 Conclusion

The protection of the environment is our responsibility, whether we are a member of the public as a citizen or consumer or as an industrialist producing consumer products, or as a part of Government, either in setting environmental policy or in regulating environmental pollution.

Concern for the environment and both national and international regulations will continue to put pressure on industries to minimize their impact on the environment. There are great business opportunities for companies to supply and adopt cleaner technologies and more environmentally friendly techniques. There are challenges to be met in developing new products and processes that are more efficient and produce less waste. There are savings to be made in avoiding the increasing costs of environmentally acceptable waste disposal. There are further opportunities. New technologies can bring together the benefits of greater efficiency, less pollution, and a minimizing of impact on the environment as a whole. Maintaining the *status quo* is not acceptable.

Environmental awareness is critical and must be central to every company's activity. It is no longer just a passing phase. The environmental impact of a process must be fully considered both for an existing process and a new process. Changes must be justified against the environmental impact. Waste minimization should be a key part of business strategy in the 1990s and the implementation of IPC has a part to play.

There is a move from a strict regulatory framework, for example as found in the planning system and water and waste regulation, through to the flexible but sophisticated approach of IPC. We are now seeing voluntary, totally integrated, environmental management, as companies appreciate the benefits of environmental commitment.

D. Slater

However, the notion of integration as expressed by the RCEP a number of years ago, embraced not only an overall view of the environment but also advocated a single regulator. It is thinking such as this which is now taking us nearer to the establishment of an environment agency. And this prospect, equally with the advent of IPC, has also produced both pressures and opportunities to forge much closer links and understandings.

© Crown Copyright 1995. Published with the permission of the Controller of Her Majesty's Stationery Office. The views expressed are those of the authors and do not, necessarily, reflect those of the Department of the Environment or any other Government department.

Sewage and Industrial Effluents

P. M. TRY AND G. J. PRICE

1 Introduction

From the time when people first gathered together to live in communities the disposal of human waste has been an environmental problem. Medieval towns and settlements were often filthy, smelly, unhealthy places, where solid and liquid wastes were commonly tipped out into the street or gutter. Although sewers had existed in Roman times, it was the widespread introduction of public water supplies since the 18th century that turned water-borne wastes into a serious health hazard. Sewage from the new water closets was discharged to road drains and so to the nearest watercourse. In the mid-eighteen hundreds, the filth and stench of the River Thames became so unbearable that the government and high society abandoned London every summer to enjoy the delights of cities such as Bath or the countryside.

The industrial revolution caused many workers to migrate to towns and cities where they lived in overcrowded slums in insanitary squalor. Waste waters from the new industrial processes also polluted watercourses, which often became open sewers. In such conditions diseases were rampant: the death rate reached 46 per thousand during the first cholera outbreak of 1832. Tuberculosis and typhoid were also endemic and even greater killers.[1] By the mid-19th century cholera was discovered to be a water-borne disease. Pollution, smells, and disease had become unacceptable and public pressure forced sewers to be constructed. These greatly improved matters by collecting sewage and carrying it away from the town gutters, streams, and rivers, but it still caused pollution where it was eventually discharged.

A Royal Commission was set up at the end of the 19th century, charged with recommending methods of treating and disposing of sewage. The Royal Commission on Sewage Disposal produced nine reports between 1901 and 1915[2] which have had a seminal influence on UK practice to this day. The Commission investigated the existing systems for the disposal of sewage, and came down firmly in favour of water-borne carriage of wastes, proposing treatment together with industrial waste water at municipal sewage works. It undertook experimental

---

[1] Sir Edwin Chadwick, 'Report on the Sanitary Condition of the Labouring Population of Great Britain', HMSO, London, 1882.
[2] Royal Commission on Sewage Disposal, Final Report, HMSO, London, 1915.

work which established the need for preliminary treatment and primary settlement, demonstrated that biological filters were more effective than contact beds, and recommended methods of treating waste water from various industries.

The eighth report recommended sewage works effluent quality limits of 20 mg l^{-1} for 5 day Biochemical Oxygen Demand (BOD_5) and 30 mg l^{-1} for Suspended Solids (SS), which have been the benchmark standards until relatively recently. But it recognized that standards should vary according to the dilution available in the receiving watercourse.

Originally sewers developed from the road drains took both sewage and surface water. Most urban areas in the UK and many other countries still have combined sewers. A major disadvantage of combined sewers is that where storm overflows are needed during times of heavy rainfall, discharges can have a severe polluting effect on the local watercourse. This is largely avoided where there are separate sewers for surface run-off and for foul sewage.

Treatment Systems

Early sewage treatment relied on land irrigation; sometimes using osier (willow) beds and treatment with lime to reduce the smells. This was land and labour intensive and ineffective on impermeable soils due to waterlogging, so alternative treatment systems gradually emerged. These involved settlement in storage tanks to greatly reduce solids discharges, and biological treatment of the settled sewage in contact beds. The beds were tanks about 1.2 m deep filled with clinker and were operated by filling with sewage and then decanting the liquid after a few hours.

Treatment processes were developed empirically, by creating conditions to speed up the natural biological breakdown of organic wastes. A hundred years ago biological filters evolved as a development from contact beds for the first time allowing effective continuous treatment. Pioneering work by Arden and Lockett[3] at Manchester in 1914 and by Haworth at Sheffield led to the development of the activated sludge process. These two processes are today still the most widely used systems for treating sewage.

Most sewage treatment plants have three or four process stages, as illustrated in Figure 1.

Preliminary Treatment

Preliminary treatment protects subsequent treatment processes from blockage or overloading and prevents damage to mechanical equipment. Large solids are either removed from the flow by screens or macerated. Inorganic grit is settled out, and the peak flow rate may be controlled.

Primary Settlement

In primary settlement most of the solid particles settle out by gravity in a settlement tank. Three types of tank can be used, *viz*.: horizontal flow, upward

---

[3] E. Arden and W. T. Lockett, 'Experiments on the Oxidation of Sewage without the Aid of Filters', *J. Soc. Chem. Ind.*, 1914, **33**, 523–539.

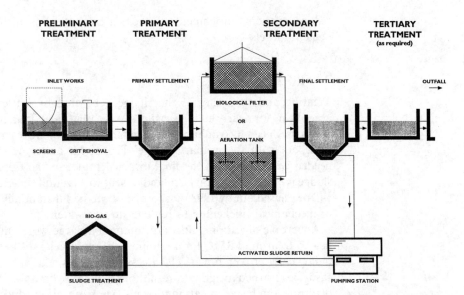

Figure 1 Diagrammatic representation of the process of sewage treatment

flow, or radial flow, depending on the size of the works, and all have outlet weir flow control. Sludge is periodically drawn off from the bottom of the tank.

Secondary Treatment

At the heart of the process is a biological treatment stage where micro-organisms oxidize the settled sewage. With biological filters the settled sewage is sprinkled over a bed of stone or slag media, which provides a habitat supporting the growth of micro-organisms. This is followed by a further settlement stage to remove the waste solid products of oxidation. In activated sludge plants (ASPs) the micro-organisms are kept in suspension in the sewage or mixed liquor, which is aerated either by bubbling air into the tank or by beating air into the surface of the liquor. Solids are again separated in a final settlement stage, and the sludge containing the active bacteria is recycled to sustain the mixed liquor.

Modern variants of filters include use of plastic media to replace the traditional media, and rotating biological contactors, in which the medium is in the form of rotating discs which are alternately submerged in sewage and then exposed to the air. There are several commonly used variants in the activated sludge process, including oxidation ditches which operate with extended aeration but without the primary settlement stage.

In recent years the use of biological aerated filters (BAFs) and reed beds has developed. The BAF is a submerged fixed-film reactor into which air is bubbled. There are similarities between the BAF and original contact beds, and the reed bed and osier beds.

Tertiary Treatment

For some plants further treatment is needed to enhance the effluent quality. This may include the removal of residual solids, ammonia, nitrate, and phosphorus to reduce nutrient enrichment of the watercourses; or disinfection to reduce bacterial levels.

Sludge Treatment

Treating and disposing of the sludge generated by settlement is expensive, accounting for nearly half the cost of sewage treatment. In many plants the sludge solids produced by oxidation in secondary treatment are returned to the inlet and settle out with the primary sludge. The extent and type of sludge treatment needed will depend on the final disposal route. In the European Union most sludge is either spread on farmland or sent to a landfill tip: sea disposal of sludge will be phased out by 1998. Disposal cost can be substantially reduced by gravity or mechanical thickening to remove surplus water.

Anaerobic digestion plants are common at large works in the UK. Sludge is heated to around 35 °C for a minimum of two weeks so that organic solids are biologically broken down. The digested sludge is relatively inoffensive and can easily be returned to agriculture, although there are limits on the acceptable levels of nitrate and heavy metals in the soils. Methane gas produced is burned to heat the sludge in the digester, and is frequently used in a combined heat and electrical power generation unit.

2 Regulatory Framework

The discharge of sewage and industrial effluents into inland and coastal waters needs to be strictly controlled to prevent environmental damage. In the UK the required quality for inland and coastal waters is achieved by the use of Environmental Quality Objectives (EQO). EQOs are site specific and take account of the use to which the waters are put, *e.g.* abstraction, fisheries, bathing. Environmental Quality Standards (EQS) are set for residual concentrations of contaminants to be discharged to the receiving waters. This approach is very different from the concept of common emission standards which is the practice in many European countries and now forms the basis of recent European legislation. The water industry in Great Britain believes that site-specific EQO and EQS is the most environmentally acceptable method of controlling discharges providing cost-effective solutions to the disposal of water-borne waste.

The principal environmental regulators for England and Wales are the National Rivers Authority (NRA) and Her Majesty's Inspectorate of Pollution (HMIP). The NRA is responsible for the regulation of the complete water cycle from abstraction to discharge and beyond. HMIP's principal responsibility is the authorization and regulation of industrial processes having the greatest potential for pollution of the environment. Both the NRA and HMIP are required to maintain public registers of applications and authorized discharge consents.

The UK Government has announced its intention to create an overall regulatory agency for environmental matters. It is intended there will be one agency for England and Wales and a separate one for Scotland. Legislation is planned to appear at the end of 1994 to enable this change to take place. In an announcement on 15 July, 1992, the Secretary of State said that the new

Environmental Agency would incorporate the existing functions of HMIP, the waste regulation responsibilities of local authorities, and all the existing functions of the NRA. The intention is that the new agency will offer a 'one stop shop' to dischargers, thus simplifying the environmental regulation framework.

Key Legislation

Over the past two decades environmental legislation in the UK has increasingly been driven by legislation introduced by the European Union. The Declaration of the Council of Ministers on the Programme of Action of the European Communities on the Environment on 22 November, 1973, reported that 'expansion cannot now be imagined in the absence of an effective campaign to combat pollution and nuisances or of an improvement in the quality of life and the protection of the environment'. The EC Environmental Policy has been pursued robustly in Brussels ever since.

EC Directives have gradually been incorporated into UK law. The significant Directives relating to discharges are the following:

(i) Quality of Bathing Waters (76/160/EEC);
(ii) Pollution caused by the Discharge of Certain Dangerous Substances to the Aquatic Environment (76/464/EEC);
(iii) Quality of Fresh Waters needed to Support Fish Life (78/659/EEC);
(iv) Protection of the Environment when Sewage Sludge is used in Agriculture (86/278/EEC); and
(v) Urban Waste Water Treatment Directive (UWWTD) (91/271/EEC).

The UWWTD is the most significant in terms of capital expenditure needed to meet its requirements for the collection and treatment of urban waste waters.

Most aspects of current water quality regulations in England and Wales are controlled by the Water Resources Act 1991 and Water Industry Act 1991. The Water Resources Act established powers for the NRA to

(i) monitor the state of the water environment, classify the quality of the waters, and set environmental quality objectives;
(ii) grant consents to discharge to water, monitor discharges, and enforce compliance with consent;
(iii) take action in the event of an offence of polluting controlled water;
(iv) maintain registers of consents to discharge and monitoring information on compliance with consent; and
(v) recover costs by a scheme of discharge consent charges.

The Water Industry Act gives sewerage undertakers the power to

(i) grant consents for the discharge of trade effluents to public sewer, monitor discharges, and enforce compliance with consent;
(ii) maintain a register of trade effluent consents to discharge; and
(iii) recover costs by a scheme of trade effluent charges.

The Water Industry Act 1991 also requires sewerage undertakers to refer applications for consents for the discharge of special category effluent (including

Table 1 Sewage connections and volumes

	Population connected to sewer/%	Population served by sewage works/%	Total no. of sewage works	No. of sewage works serving >100 000
Belgium	58	25	292	13
Denmark	94	92	1805	18
France	65	50	7805	—
Germany (West)	91	86	8456	137
Greece	40	10	26	2
Ireland	66	25	530	2
Italy	—	—	3783	102
Luxembourg	96	76	324	1
Netherlands	92	88	485	64
Portugal	38	37	166	—
Spain	80	43	1595	—
UK	**96**	**83**	**7645**	**102**
England and Wales	96	83	6524	—

those from a prescribed process or containing prescribed substances) to the Secretary of State for consideration by HMIP.

The Environmental Protection Act 1990 gives HMIP the powers to control 'prescribed processes' as specified in the Act. The aim of the Act is to reduce pollution by certain dangerous substances, especially List I and List II,[4] of the atmosphere, watercourses, and land.

3 Sewage Treatment

Water-borne Discharges

Treated sewage is over 99.9% water and contains a wide variety of substances in small or trace concentrations. 96% of the UK's 55 million population is connected to public sewers and 83% of this is served by sewage treatment works. The remaining 17% discharges to sea through marine outfalls, often without even preliminary treatment. Table 1[5] shows the European position and shows the UK as one of the most advanced countries.

Discharges to inland waters are controlled mainly by the NRA to prevent overloading the receiving waters with organic polluting matter. This maintains dissolved oxygen levels and protects the flora and fauna which often thrive only within a narrow range of environmental conditions. In severe cases of pollution the result could be no aerobic life at all. For discharges to oxygen rich marine waters the biochemical factors are not so important and bacterial content can be the limiting criterion.

The majority of Water Resources Act consents are long standing many having

---

[4] 'Dangerous Substances in Water: A Practical Guide', First Edition, Environmental Data Services Ltd., 1992.
[5] 'Water Facts 1993', Water Services Association, 1993.

first originated from the Rivers (Prevention of Pollution) Acts of 1951 and 1961. The standard requirement for most sewage effluent is for a limit on five day BOD and Suspended Solids (SS) with a lesser number of sensitive discharges requiring a limit on Ammoniacal Nitrogen (AmmN). The basic treatment requirement for most inland works, therefore, is primary settlement, a biological treatment stage, and secondary settlement. The degree of treatment necessary in the second stage will depend upon whether an ammoniacal nitrogen limit has been applied. The values of quality parameters are determined by the NRA after assessing the volume of discharge and quality objectives for the receiving watercourse, taking into account the dilution available.

The percentage of sewage works in England and Wales failing to comply with consent conditions improved from 23% in 1986 to 6% in 1991. The UK water industry has spent around £11 billion over the period 1989–93 since privatization and is currently committed to further improvements.

The formats of consents to incorporate the requirements of the UWWTD and the Water Resources Act have recently been drafted. Provision of treatment facilities under the UWWTD will be more demanding than under current UK law, especially for sensitive waters and coastal discharges. Standard Waste Water Treatment Directive consents in the UK for works treating a population equivalent (p.e.) of >2000 will require reductions in BOD of 70% and Chemical Oxygen Demand (COD) of 75%, or fixed emission standards of 25 mg l^{-1} BOD and 125 mg l^{-1} COD. The optional Suspended Solids limit of 35 mg l^{-1} or reduction of 90% is not being adopted in the UK. Works treating <2000 p.e. should have treatment appropriate for local conditions.

Discharges to 'sensitive' waters will have additional requirements for phosphorus and nitrogen, since excessive levels could stimulate plant and algal growth and alter the ecology of the water. A phosphorus limit of between 1–2 mg l^{-1}, depending upon the volume of discharge, or a reduction of 80% will be required. Total nitrogen limits of 10–15 mg l^{-1} N, or a reduction of 70–80%, may also be necessary. Discharges to esturial or coastal waters in 'high natural dispersive areas' will only require primary treatment with a reduction in BOD of at least 20% and a reduction in suspended solids of at least 50%.

To comply with the new standards, secondary treatment will have to be provided except for some small works and in areas of high natural dispersion. These areas have been agreed with the NRA, but comprehensive studies will have to be undertaken to prove that there will be no detrimental effect in not providing secondary treatment. The introduction of sensitive areas will also result in the provision of a nutrient-stripping tertiary treatment stage. The eastern side of the country is most affected since the rainfall is less and dilution in rivers is subsequently lower.

In the UK, most agglomerates of >2000 p.e. are already served by a sewerage system. However, as most of these sewers are combined, intermittent storm discharges can adversely affect the receiving watercourses. Under the UWWTD remedial work must be undertaken to unsatisfactory overflows by provision of additional sewer capacity and/or screening to remove aesthetic pollution by plastics and other debris from the watercourses.

The provision of sewage treatment works for esturial and coastal populations, which currently may have no treatment, will mean high capital expenditure and increased operational costs. Capital expenditure and increased operational costs will also result from the provision of nutrient removal and storm overflow improvements.

The EC Dangerous Substances Directive and the related UK regulations, will impose much tighter control on prescribed substances discharged in sewage treatment works effluents, to prevent toxic substances from entering the aquatic environment. Strict trade effluent control at industrial premises is the only economic approach to this problem.

Sludge

Sewage sludge is the mainly organic residue remaining after sewage treatment. Sewage originates from a wide diversity of households and industrial activity and so sludge constituents can vary considerably over a short timescale.

The sludge is usually treated before disposal by a combination of the following processes:

(i) gravity or mechanical thickening to reduce the volume of the liquid sludges;
(ii) anaerobic digestion to reduce the overall weight of sludge solids and reduce smell and pathogen levels;
(iii) mechanical dewatering to produce a stable solids product; and
(iv) drying by heat treatment to produce a granular or pelleted product, which is odourless and pasteurized to control bacterial levels.

Disposal methods must be economical and minimize adverse environmental consequences. Where possible, the beneficial characteristics of sludge can make recycling a practical alternative. Table 2 compares some of the qualities of sewage sludge.

Sludge Disposal Routes

Disposal routes are strictly controlled by EC legislation and UK Regulations to maximize the benefits of sludge and prevent environmental damage. Table 3 shows the quantities of sludge and disposal routes in 1991.[6]

The UK Government agreed at the Third North Sea Conference in 1990 that sea disposal would cease after 1998; this is also a condition of the UWWTD. The additional sewage treatment arising from the UWWTD will result in a significant increase in sludge production, and the predicted quantities of sludge and disposal routes are also shown in Table 3. An EC Directive (86/278/EEC) aims to maximize the benefits of sewage sludge within the community whilst ensuring the potential adverse effects are carefully monitored and controlled. Some of the benefits of alternative disposal routes are discussed below.

Agriculture

This is the most common option. Liquid digested sludges are usually spread onto the surface of grasslands. Care has to be taken to prevent run-off from polluting

---

[6] 'UK Sewage Sludge Survey', DoE, 1993.

Table 2 Sewage sludge properties

Beneficial
  (i) The water content varies from 99% down to 5%, dependent upon the amount of thickening or dewatering which takes place, and can be of benefit to a growing crop.
 (ii) Provision of some essential trace elements for crop growth.
(iii) Nitrogen and phosphorus, at levels between 3% and 6% of the dry matter present, are valuable crop nutrients.
(iv) Organic matter, which varies between 50% and 80% of the dry matter present, improves the soil structure and its ability to retain moisture.

Non-beneficial
  (i) Pathogen levels are high in untreated sludges but can be substantially reduced, dependent on the type of sludge treatment.
 (ii) Untreated sludges smell badly and can be the cause of many complaints; dependent upon the treatment process selected, the smell problem can be reduced or even eliminated.
(iii) High levels of fats and greases cause sludges spread onto pastures to cling to the sward rather than being easily washed down to the root systems.
(iv) Large amounts of paper and plastics are unsightly and if spread to pastures may be ingested by animals.
 (v) Contaminant levels, *e.g.* high metal content, may preclude safe disposal to farmland.

Table 3 Sludge disposal in the UK

Disposal	1991 tds*	Estimated 1999 tds
Agriculture	465 000	777 000
Dedicated	25 000	30 000
Sea Disposal	334 000	0
Incineration	77 000	382 000
Landfill	88 000	35 000
Beneficial	68 000	137 000
Within Curtilage	50 000	91 000
Uncertain	—	294 000
Total	1 107 000	1 746 000

*tds = tonnes dry solid

surface or groundwater sources. A typical application rate is about 100 m^3 ha^{-1} annum^{-1}. The high levels of ammoniacal nitrogen in digested sludge ensure that grasses grow rapidly.

Untreated liquid sludges are usually either surface applied to arable land, which is then ploughed, or they are injected beneath the surface. Incorporation into the soil prevents the spread of pathogens and keeps smell problems to a minimum. Nitrogen release is much slower, allowing it to be taken up by an arable crop. 'Cake' sludges, which are often untreated, are generally ploughed into arable land at a rate of approximately 50 tonnes ha^{-1} annum^{-1}.

Within England and Wales the Directive is enacted under Regulation[7] and further enhanced with a Code of Practice.[8] These regulations require sludge producers usually Water Services Companies, to analyse receiving soils to a depth of 25 cm before sludge is applied to ensure that the following conditions are satisfied:

(i) the pH is >5.0, thus minimizing heavy metal transfer from soil to plants;
(ii) the concentrations of each of seven heavy metals (Zn, Cu, Ni, Cr, Pb, Cd, Hg) are less than prescribed limits.

They also require analysis of sludge quality to ensure that

(i) this metal loading of the seven heavy metals must not exceed prescribed limits;
(ii) the concentrations of lead and fluorine in sludges destined for grassland application must not be higher than prescribed levels; and
(iii) the concentrations and quantities of nitrogen and phosphorus are established.

This information must be recorded in a register for audit and inspection and is used to demonstrate appropriate environmental control. Information about soils and sludge application must be made available to the land owner.

The associated Code of Practice further requires that

(i) the soil is again sampled, at a depth of either 15 or 7.5 cm for pasture or arable purposes, respectively;
(ii) the concentrations of each of the seven heavy metals in the sample are less than prescribed limits;
(iii) only treated sludges may be placed on grassland and animals must not be allowed to graze there for three weeks after sludge application in order to reduce the risk of infection; and
(iv) 'cake' sludges may not be used on grassland for fear of ingestion by animals.

In order to minimize the impact of excess nitrogen on the environment and prevent high levels in water, another Code of Practice[9] seeks to:

(i) limit the amount of organic nitrogen applied to land in any one year to 250 kg ha^{-1};
(ii) ensure that it is applied only when there will be no fear of run-off to watercourses; and
(iii) ensure that the nutrients are applied when the growing crop can use them.

Further constraints on sludge recycling to farmland include the following:

(i) the requirement to apply no sludge to land constrained by the 'set aside' scheme of the Ministry of Agriculture, Fisheries and Foods (MAFF); and
(ii) the need to add little or no sludge to land designated as 'Environmentally Sensitive'. This scheme compensates farmers who limit the amount of

---

[7] 'The Sludge (Use in Agriculture) Regulations 1989', DoE, 1989.
[8] 'Code of Practice for Agricultural Use of Sewage Sludge', DoE, 1989.
[9] 'The Code of Good Agricultural Practice for the Protection of Water', DoE, 1991.

nutrients added and thus preserve, rather than enhance, the fertility of designated land.

In addition to agricultural use, some sludges are recycled as top dressings for tips, road embankments, and for recreational areas, such as golf courses.

Landfill

Normally landfill sites are used when farmland lies outside an economic radius for sludge recycling or when the sludge is contaminated and poses a pollution risk to the soil. Many 'cake' sludges are disposed to landfill since filter pressing significantly reduces both the volume to be transported and the reception fee charged at the landfill. The sewage treatment preliminary processes and sludge screening produces rags, plastics, wood, and grit. These materials are usually also sent to landfill sites for disposal.

The Environmental Protection Act 1990 requires that Landfill sites are registered by county based Waste Regulatory Authorities. These Authorities set strict limits and controls on the quantity and type of waste materials which can be accepted into registered sites. These limits are specific to a particular site and are designed to ensure that the materials are compatible and will be retained and treated within the site without environmental pollution.

Forestry

Sludge application in forestry is difficult. Land is often of poor quality and far from the towns which produce the sludge; it may be steep and difficult to incorporate sensible amounts of the sludge without causing run-off. Access may be restricted by inadequate road systems or even by heavy tree growth itself before the saplings are thinned.

Disposal to forestry is now covered by a Code of Practice.[10] This code recognizes the tree growth benefits of adding nutrient and organic matter to land often of poor quality. It also seeks to ensure forest land is subject to many of the precautionary constraints which apply to sludge recycling on farmland; thus nutrient and metal loadings are controlled in a similar fashion.

Sea Disposal

Until 1998 sea disposal will remain under the careful control of MAFF, under the Food and Environmental Protection Act 1985. Disposal sites are chosen carefully and licensed by the Ministry only after passing an exhaustive checklist designed to ensure that no harm will result to the marine environment. The sludge producer must discharge the sludge at specific states of the tide and install equipment which continuously monitors the whereabouts of the vessel, indicating whether it is travelling full or empty and whether it is carrying sludge or sea water ballast. The licence limits the sludge volumes and the maximum loads of heavy

---

[10] 'Manual of Good Practice for the use of Sewage Sludge in Forestry', HMSO, London, 1992.

metals as well as the concentrations and loads of PCBs and some organochlorine pesticides.

Incineration

The closing of the sea disposal route has forced Water Service Companies to develop new disposal strategies. Farmland disposal may not be an option for coastal cities producing large quantities of sludges, perhaps contaminated by industry. Incineration requires sludge to be dewatered to at least 30% dry matter. HMIP strictly controls the emission levels of metals, particulates, and various gases such as sulfur dioxide and carbon monoxide, and by-products of combustion, *e.g.* dioxins, under the Environmental Protection Act 1990. Flue gas scrubbing can minimize atmospheric emissions. Some 30% of the sludge dry matter remains after combustion as ash and this normally is removed to landfill.

Atmospheric Emissions

Sewage treatment can be an odourous process under certain anaerobic conditions. The sewerage systems and treatment works are usually designed to minimize the production of odours, which are usually sulfur based. Hydrogen sulfide odours from sewerage systems and primary processes occur in septic conditions caused by long retention periods, hot weather, and very strong sewages. Hydrogen sulfide and mercaptans can be released during desludging of primary tanks or from areas of turbulent flow if the sewage has turned septic. Hydrogen sulfide is a toxic gas and in concentrations above 10 p.p.m. can be lethal to workers in sewers and pump sumps. It is also corrosive to concrete and equipment. Treatment units and problem areas can be covered and the odorous gases treated either chemically or biologically.

A beneficial gas, methane, is produced from the anaerobic digestion of sludges. Power generation is generally considered economic for population equivalents $>100\,000$ and sometimes for plant of $>50\,000$ p.e. The electricity can be sold to the Electricity Industry at a premium rate under non-fossil fuel agreements. Excess methane, which is a greenhouse gas, is either flared off or released to the atmosphere. However, $<3\%$ of the UK's estimated methane emissions originate from sewage treatment.

Public Health

Micro-organisms are a ubiquitous and intimate part of our daily existence. Every day large numbers of them are discharged into the wastewater system from healthy and unhealthy people through diverse human activities. Risks arising from the wastewater generated will depend on the general state of health of the population. Inadequate hygiene and sanitation systems can cause contamination of foods and water used for drinking. Epidemics of cholera and typhoid[11] can occur and spread rapidly, particularly in poor and deprived areas. The South

---

[11] P. West, 'The ecology and survival of *Vibrio cholera* in natural aquatic environments', in 'Cholera update', *PHLS Microbiol. Digester*, 1992, **9**, 13–42.

Table 4 Quality of bathing water 1989–92, % compliance with CEC directive

Sea Water	1989	1990	1991	1992
Belgium	94	90	85	90
Denmark	—	—	93	95
Germany	—	—	64	73
Greece	—	—	90	97
Spain	80	—	88	93
France	84	—	87	79
Ireland	96	—	97	94
Italy	84	—	90	92
Luxembourg	—	—	—	—
Netherlands	—	—	86	86
Portugal	—	87	83	83
UK	**76**	**77**	**76**	**79**

American pandemic of 1991 and Rwanda of 1994 are recent illustrations. Hostilities such as the Gulf War can often result in the breakdown of sanitary systems and increase risk of water-borne disease.

Wastewater treatment has generally been directed towards the reduction of aesthetic and obvious polluting aspects of wastewater discharges, rather than microbiological aspects of the processes. Most treatments achieve some reductions in faecal, including pathogenic, micro-organisms and some plants may be augmented with various tertiary treatments including lagooning, sand filtration, membrane filtration, and disinfection which further reduce bacterial levels. Tertiary treatment or chemical disinfection are used particularly where microbiological standards apply in the receiving waters. These standards tend to be most depending where recreational use is to be made of the receiving water, as with inland waterways and resort bathing waters.

Microbiological standards relating to coastal and esturial waters include European directives on bathing water quality and the water quality in shellfisheries. Public and media concern about sewage disposal in these areas has related not just to compliance with these standards but also to more obvious aesthetic signs of pollution, including material washed up on beaches or floating in or on the water. This has an adverse psychological effect on the public well-being and recreational value, quite apart from any subsequent morbidity associated with faecal contamination of the water. The 'Good Beach' guide statistics and 'Blue Flag' awards include standards relating to cleanliness and provision of facilities such as toilets at the beach site, in addition to the achievement of microbiological quality standards. European compliance with the Bathing Water Directive is given in Table 4. Gradual improvement in the UK in bathing water quality is evident.

In fact, the microbiological standards of the Bathing Water Directive were set arbitrarily and not on the basis of knowledge of public health significance. Subsequent epidemiological studies aiming to relate standards to health criteria have been widely criticized by scientists on statistical grounds.[12] In practice,

[12] R. Philipp, 'Risk assessment and microbiological hazards associated with recreational water sports'. *Rev. Med. Microbiol.*, 1991, **2**, 208–214.

these studies are both difficult and expensive to perform. The most recent and largest study of this kind[13] concluded that there was a correlation between faecal micro-organism numbers (particularly Enterovirus) and gastro-intestinal symptoms. The correlation in those who had been exposed to the water was only significant, however, when counts significantly exceeded the mandatory, imperative standards of the Bathing Water Directive. Other categories of symptom such as eye, ear, nose, throat, and skin disorders were correlated with degree of water contact, duration, and intimacy of exposure, rather than concentrations of microbial indicators of faecal pollution. The Bathing Water Directive microbiological standards are being reviewed and amended, giving greater importance to faecal Streptococci in line with experience, although European consensus has yet to be achieved.

Viruses are of topical interest in relation to exposure risks from wastewater use and discharged. Public Health Laboratory figures show that Hepatitis A, responsible for infectious jaundice, has increased since 1987.[14] Shellfish consumption can increase risk but no link has yet been made with recreational water exposure. There is no evidence supporting risk to the public from exposure to recreational waters or sewage from HIV, the causative agent of AIDS.

Bathing waters can suffer from potential microbiological hazards other than from sewage discharge. Notably, environmental contamination can result from gulls which have a prevalence of enteric bacterial pathogens such as *Salmonella* and *Campylobacter* in their faeces.[15]

The position is similar where inland waters are used for recreation. The concentration of faecal organisms is the principal factor in the risk of disease. Recreational exposure, including canoeing, swimming, sailing, and fishing differs only slightly from marine waters in the degree and intimacy of exposure. Rivers act as repositories for large bodies of treated wastewater, surface water, and agricultural run-off. Sewage effluents, although subject to stringent discharge consents, still have significant inputs of faecally derived microbes. Abstraction of river water for drinking purposes places constraints on the location of the water purification works and the type of treatment required.

The often quoted risks of Leptospirosis, or Weil's disease, must be kept in context. Sewage workers may have an increased risk from indirect exposure to rats and this has led to confusion with the risk of exposure to sewage. In fact the disease is usually contracted by exposure to the urine of infected rats or of urine-contaminated stagnant water. Infection occurs through cuts and abrasions or via the mucous membranes.[16]

---

[13] E. B. Pike, Health Effects of Sea Bathing (WMI 9021)—Phase III Final Report to the Department of the Environment, 1992, Report No. DoE, 3412 (P).

[14] PHLS Working Group, 'The present state of hepatitis A infection in England and Wales', *PHLS Microbiol. Digest*, 1991, **8**, 122–126.

[15] D. J. Gould, 'Gull droppings and their effects on water quality', Water Research Centre Tech. Report 37, 1977.

[16] S. Moore, 'Occupational exposure to Leptospirosis. Reducing the risk of Weil's disease', *Occupational Hlth. Rev.*, Dec. 1991/Jan. 1992, 30–32.

Environmental Impact Assessment

Sewage treatment works have an environmental impact which must be considered in the selection of the site location. This is especially true for new works on green field sites, such as many of the coastal sites required under the UWWTD.

Developments with a potential for environmental impact are covered by the EC directive 85/337/EEC concerning the effects of certain Public and Private projects on the environment. The directive was implemented in the UK through the Town and Country Planning (Assessment of Environmental Effects) Regulations in 1988 and amended in 1990 and 1994. Within the UK, planning applications for sewage treatment works are determined by County Planning Authorities under their responsibility for waste disposal or the Local Planning Authority. Sewers and small works extensions constitute 'permitted development' under the Town and Country General Development Order[17] and are not subject to Environmental Assessment (EA). The EA of a new sewage works should cover associated infrastructure development such as roads, sewers, discharge pipes, and power supplies. EA is a process which enables the developer and planning authorities to assess the various effects leading to the selection of a preferred treatment works process and location. It is summarized in the Environmental Statement (ES) which accompanies the Planning Application. In the UK, the developer is responsible for both the scope and the quality of the Environmental Statement. It is good practice to scope the EA in consultation with the statutory bodies such as the NRA, English Nature, the Countryside Commission, English Heritage, the Planning authorities, and the local community.

In the EA of sewage and sludge treatment, consideration is given to both significant operational and construction effects. The ES is a public document and offers an opportunity to describe how any potential negative effects will be resolved.

Indirect effects and distant effects are also relevant. In sewage treatment, energy use may give rise to greater greenhouse gas production from the sludge. Transport of solid waste will given rise to direct traffic impacts, both local and possibly distant, and indirect effects through fuel use (NO_x, CO_2, and particulate emissions). Effluent discharges also tend to have distant effects in terms of downstream water quality, which will be considered in the consenting process, but indirect water quality effects such as the risk of overflow from sewers due to stormflows, blockages, or pump failure also arise.

Site Selection

The primary aims of site selection for a waste management site are[18] to ensure that the site conforms with the project specification and that the project is acceptable to the local community. This must be balanced with minimizing environmental impact and development costs.

For sewage treatment works, the discharge is usually the key environmental

---

[17] Town and Country Planning General Development Order, (SI No. 1813), 1988.
[18] J. Petts and G. Eduljee, 'Environmental Impact Assessment for Waste Treatment and Disposal Facilities', John Wiley and Sons, Chichester, 1994.

Table 5 Typical constraints in sewage treatment site selection

	Engineering	Environmental
Topography	*Altitude*–dependent on catchment and discharge heights *Slope*–dependent on process choice	*Visual*–low quality landscape or secluded
Population	Not within 400 m of residential development. Close to 'A' road for transport of sludge. Short sewage transfer time	Not protected heritage site. Not protected conservation site. Not area of outstanding natural beauty.
Receiving Waters	Close to receiving water. Not within flood plain. With acceptable storm overflow route.	Avoid: bathing waters, potable abstraction, fisheries (salmonid), shellfishery, sensitive areas (E), SSSIs
Socio-economic	Low land value (purchase price)	Low recreational value

effect and therefore the location of the receiving water is a prime consideration. The site position in relation to the sewers is similarly critical. The process choice, discharge quality, and location, are therefore interrelated. Best results are achieved when water quality monitoring and modelling required for the consenting purposes is undertaken within the EA.

In order to identify possible sites, an overlay technique can be used. The engineering and environmental constraints, which may be exclusionary or preferred, are specified (see Table 5).

Future land use and planning designations can be taken from local and county structure plans. The areas delimited by the constraints are drawn onto maps and overlaid. The resulting map will show areas most suitable for the sewage works construction. In practice, the site selection process may well rely on some compromise. For coastal towns the topography and earlier marine treatment concepts mean that the sewage usually drains through congested urban areas to the coast. The construction of the required sewage treatment facilities in urban locations poses significant environmental and engineering problems.

However, plants can be designed and constructed so that they blend into the local urban environment. Perhaps the biggest problem is to get the public to accept that a traditional wastewater treatment plant can become a 'purification centre' without the negative image that existing plants sometimes generate. Irrespective of whether plants are built above or below ground their outward appearance can completely disguise what is happening inside and, by careful design, no external smells or noise should occur. Careful choice of use of space above the plant can help offset high urban land costs. Building enclosed works is very expensive but so too is the construction of sewers and associated works needed to transfer sewage from the coast to a suitable inland site for treatment and return back to the coast.

Process Selection

The process selection cannot be entirely separated from the site selection. Once the potential sites are identified, and the likely discharge quality known, then the optimum process can be established for each site. Where landscape or odour and noise nuisance or land availability are significant considerations, compact treatment methods including biological aerated filtration (BAF), enhanced settlement, or high rate activated sludge offer benefits for covering and underground installation. However, for rural sites, conventional treatment offers energy efficiency and both cost and risk benefits. Reedbeds and lagoons, which are used more extensively in the USA and Europe, have low visual impact and are suitable in rural landscapes.

4 Industrial Effluents

Discharges to Sewer

Industrial effluent control is a highly regulated field. Regulation has driven considerable developments to minimize the impact of industrial waste on the environment. The principal legislation controlling industrial discharges to sewer is the Water Industry Act 1991. Trade effluent discharged to the public foul sewer must have the prior authorization of the Sewerage Undertaker in the form of a consent. This will include quantity and quality conditions and provides the framework of control.

Industrial effluents are inherently complex and variable. Effluents discharging into the sewerage system, alone or in combination, can result in gaseous emissions or blockages. Where the waste contains sulfur-containing organic compounds or sulfates, the activity of bacteria can form sulfides and other malodorous compounds. This occurs in anaerobic conditions such as rising mains or large low-turbulent gravity sewers and may lead to the formation and release of methane gas. The disposal of solvents is strictly controlled at source to prevent the formation of vapours, and the discharge of petrol to sewers is specifically prohibited. Accidental discharge of either could cause explosive atmospheres.

Industrial effluents need to be strictly controlled to ensure the build up of solids or deposition of grease does not cause a blockage which would result in the premature operation of stormwater overflows. The discharge of organic matter can significantly effect the ecological balance of the watercourse. The possible presence of toxic substances can also exacerbate the problem, although this should have been taken into consideration when setting the conditions on the sewer discharge consents.

Heavy metals such as zinc, copper, and lead are normally found in domestic sewage. Significantly higher concentrations of these and other metals are often found in industrial wastes. The majority of these metals end up in the resulting sewage sludge leading to problems if it is to be recycled to land. In all cases the industrial content of the sewage can substantially effect the final concentrations of toxic substances in the sludge for disposal. Control of metal bearing effluents at source is the prime method of reduction or elimination. This may require the industrialist to install an extensive pretreatment plant, such as metal precipitation, or the total containment of certain noxious wastes for disposal by a specialized contractor.

Certain industrial effluents can contain particularly persistent or toxic compounds such as cadmium, mercury, and persistent pesticides such as pentachlorophenol or γ-hexachlorocyclohexane. Some of these organic chemicals such as pentachlorophenol may undergo some biological degradation through the treatment process. A portion, however, will remain unaffected and retain its pesticidal properties. Its discharge into the receiving environment must be minimized at source. Release of pollutants must be prevented where possible or minimized, and all emissions must be rendered harmless to the environment. The controlling authority in England and Wales for these 'special category' substances is HMIP which applies the concept of Integrated Pollution Control (IPC) by the means of Best Practicable Environmental Option (BPEO) (see Chapter 1). Many industrial processes result in the release of pollutants to a variety of receiving media and, under IPC, detailed limits and operating standards are imposed to control the range of emissions.

Some of the constituents of industrial effluents will remain in the liquid phase and may be discharged as part of the sewage works effluent. The NRA will set appropriate limits in the consent to protect the quality of the watercouse. Again in order to minimize the impact on the environment, certain industries may need targeting to reduce their inputs.

Discharges to Watercourses

Some industries opt to discharge direct to watercourses rather than to the public sewer system. Such discharges are controlled by the NRA who grant a consent to discharge similar to that which a Sewerage Undertaker grants to a sewer discharge. Conditions will be imposed which relate to the receiving watercourse using the same criteria used to determine the consent levels on a sewage works final effluent. Compared to a public foul sewer discharge, the conditions on the consent will be much tighter in view of the potentially immediate affect on the receiving water. The risk of failure is ever present and most industrialists prefer the option of a discharge to the sewage works. The sewage works, however, can be severely affected by a sudden strong industrial discharge. The treatment process may be inhibited with a deterioration in the final effluent quality and harm to the environment.

Direct treatment by the industrialist may prove difficult if there is an imbalance of available nutrients for conventional biological breakdown. The waste streams may lack nutrients such as nitrogen and phosphorus which will restrict biological activity. Mixing with domestic sewage at a sewage works will restore nutrient balance and allow the waste to be adequately treated. In these cases sewer disposal may be the best option. However, for some food manufacturing sites it may be preferable to discharge to sewer at the earliest opportunity and retain the site solely for food production to minimize any possible contamination of their product.

Other Waste Disposal

Whatever treatment process is chosen, the plant will inevitably produce other wastes for disposal to the environment in addition to the final liquid effluent. This can be solid waste which may be sent to landfill, such as pressed sludge, or a liquid sludge for off site disposal. Both are likely to involve haulage or tankering, with similar restrictions on disposal to those applied to sewage sludge. Metallic or complex organic compounds can be an environmental issue and must be considered when deciding on the optimum disposal route for the sludge. As before, BPEO will be applicable in such circumstances.

In some cases a solid or liquid waste produced as a consequence of an industrial process or effluent treatment will be designated a controlled waste. This type of waste will be subject to the duty of care laid down by the Environmental Protection Act 1990, the so-called 'cradle to the grave principle'. This ensures the security of the life history of a waste from the point of origin to the point of final disposal. This duty applies to the producer of the waste and requires that reasonable measures are taken to prevent the unauthorized or harmful deposit, treatment, or disposal of the waste and to prevent the escape of waste from his or any other person's control. In addition, the producer must ensure that the transfer of waste is made only by an authorized person and with appropriate documentation.

Effect of UWWTD

The UWWTD will have a significant impact on industrial effluent disposal. The introduction of a biological secondary treatment stage in addition to the solids separation primary stage at some esturial and coastal sewage works will significantly increase industrial sewage disposal costs. The Directive also requires that certain industrial effluents which discharge directly to the aquatic environment will have to conform to particular standards which will be determined on an individual case basis by the appropriate regulatory authority. This applies to biologically degradable wastes with a population equivalent of >4000. These industries, chiefly the food processing sector, will need to extensively review their current effluent production and disposal facilities. In future, the implementation of waste reduction and effluent management programmes, both for direct river discharges and sewer disposal, are inevitable to further reduce the impact on the environment.

5 Future Developments

Factors Influencing Direction

The improvements and new technologies that will emerge for sewage and industrial effluent treatment into the next century will largely be determined by the driving forces influencing the industries over the next few years. In the European Union and much of the developed world the main factors are legislation, costs, and public image. In addition, land area requirements or water re-use are strong influences in some countries where space is at a premium (such as Japan) or water supply is limited (as in Australia and parts of the USA). In developing countries the main driving forces influencing development of sewage treatment are usually costs, the need for basic health care, and lack of water.

The implementation of legislation in some third world countries is often haphazard, with relatively little regard for public concerns. There is, however, a desperate need for appropriate basic low-cost sewage treatment, much of which is already available if it can be afforded and has the commitment of governments.

In the industrialized nations public image and perception often leads to new legislation. This, in turn, leads to costs in implementation which then create a new public image. This is clearly the case in the UK where concern over environmental factors has been partly replaced by concern over the cost to the public in implementation. However, the trend towards higher standards of treatment whilst making the most efficient usage of existing assets will inevitably continue.

Any new technologies and process improvements will be influenced by a variety of factors including climate, geography, and—more importantly—the investment in existing assets. Improvements to, or optimization of, existing processes will be dominant as there are relatively few instances where complete new treatment works will be needed. When limited areas of land are available in urban locations there will have to be a choice between pumping the wastewater inland to a conventional treatment works or constructing a covered urban installation. The implementation of either type of solution will be a major influence on the locality and it will be essential to initiate early full environmental assessments together with full public consultation.

In both Europe and Japan there are examples where local circumstances have led to the provision of office buildings, car parks, gardens, or sporting facilities as part of urban sewage treatment works. In some Japanese cities new buildings over a certain size must now incorporate sewage treatment/greywater recycling systems.

Although there is sufficient knowledge to construct and operate covered or underground installations safely, and to prevent noise and odour nuisance, it is unlikely that sewage treatment will readily lose the negative image which originates with the existing traditional plants. The possibility of constructing properly designed treatment plants in the midst of residential areas will require a comprehensive change in public opinion over may years. In larger communities it may be that a better alternative to a town centre development would be to pump the sewage to a suburban industrial area or manufacturing zone where the provision of a treatment works as a single-storey 'purification-factory' would be more acceptable. An additional advantage of this alternative is that the few town-centre sites which are available remain for use as less controversial underground storm water settlement and attenuation tanks which will also be needed in the future. Fortunately, during the last decade new treatment methods have been developed which will enable secondary treatment to be provided on very much smaller sites than traditional processes. Table 6 compares relative land

Table 6 Relative land areas required for sewage treatment

Process	Nitrifying/%	Non-nitrifying/%
Biological filters	100	50
Activated Sludge	12	7
Biological Aerated Filter	2	1
Membrane Technology	6	5
Chemical Dosing, *etc.*	80	40

area requirement for different treatment processes. Extra treatment capacity and land is required for works which nitrify the ammonia content than for those performing only carbonaceous oxidation.

Process Optimization

As has been discussed earlier, there is a wide range of different components to be found at treatment works. With this comes an even greater number of ways of operating them. In order to optimize the operation of treatment processes and select the optimum operating conditions for individual process components, increasing use of computer modelling and process simulation will be made. The use of computer software in process simulation of works performance allows the optimization of the operation of the whole works to give higher standards of treatment and more efficient usage of existing assets.[19] Such use of software makes it relatively easy to evaluate performance under different conditions without the expense or risk associated with running real trials on a functioning works. Additionally, computer modelling of specific components such as settlement tanks and aeration lanes will lead to modifications in design and improvements in treatment unit performance.[20]

In conjunction with modelling and practical trials, a number of broad strategies will be examined to improve commercial performance in the water industry and these are given in Table 7.

Process Integration

Where feasible, the integration of sewage treatment processes with other industries offers significant advantages. The siting of a power station near industrial processing units and a sewage works allows good quality effluent to be provided for power station cooling water which, in turn, supplies excess heat to industrial and food processing units; these, in turn, produce effluent for treatment. When energy from biogas production and soil conditioners/fertilizers

---

[19] L. Stokes, I. Takacs, B. Watson, and J. B. Watts, 'Dynamic Modelling of an ASP Sewage Works—A Case Study', Proceedings of the 6th IAWQ Workshop on 'Instrumentation, Control and Automation of Water and Wastewater Treatment and Transport Systems', ed. B. Jank, IAWQ, Burlington, Canada, 1993, p. 105.

[20] S. M. Lo, M. Hannan, N. Hallas, I. P. Jones, and N. S. Wilkes, 'Multi-Phase CFD Applications in the Process Industry', Proceedings of the 'World User Association of CFD' Conference, Basel, May 1994.

Table 7 Process optimization

	Flow/load balancing	Work in the USA indicates that up to 15% additional treatment capacity could be achieved.
	On-line monitoring	Diversion of incoming feed to balancing tanks when monitors indicate to control shock and peak loads. Diversion of effluent for 'polishing' when instrumentation indicates it to be required.
	Provide load management within works	Avoid operationally generated shock loads, provide small separate treatment devices within the works to contain shock loads and protect the 'core' process, *e.g.* mini-filter on supernatant return.
	Pre-treatment in sewers/rising mains	Dependent on appropriate locations. Modern instrumentation offers scope for additional control both within the 'treatment pipe' and within the catchment.
	Chemical dosing	Greater use of chemical treatment to up-rate primary treatment and thus decrease loadings on secondary treatment. Economic analysis will be used to determine if higher running costs are offset by capital savings. Use of dosing to control peak loadings to works.
	Review loading rates	Analysis indicates standard design parameters are frequently conservative compared to actual results. New loading rates should reflect the required method of monitoring. Settlement rates can be enhanced by lamella type devices or by chemical dosing, but this must be accompanied by a review of desludging and sludge thickening practice. Biochemical loading rates could be enhanced by better distribution, insulation and weather protection of units, recirculation of effluent, and flow balancing.
	Review traditional design and operational practice	The enhancement of existing assets will include improved media in existing biological filters, improved larger aerators in existing tanks, revised operating sequences, and alternative desludging regimes.
	Improve 'critical' facets of treatment processes	Examine the benefits of covering biological filters, pumped recirculation, regular and controlled filter 'flushing', and settlement enhancement devices
	Tighter trade control	Minimize input of difficult wastes to the works. Improve control of peak loads to works. Greater use of on-site treatment to deal with low volume high strength trade waste at source as an alternative to extension of works.

from sludge drying processes are added, not only are treatment costs subsidized, but the integrated 'environmentally sound' approach will improve public image.

New Technologies

To date, the introduction of new technologies into sewage treatment has been slow. In part this is due to the conservative nature of an industry where processes are expected to last in excess of 20 years. This, coupled to the high cost of introducing new technology to the treatment of a low value product has lead to few truly new systems being introduced. However, as costs for new technology reduce and the need to improve treatment and process economics increases, this situation will change. Existing processes have been highlighted earlier, but a few of the new techniques that will influence future developments are mentioned here.

On-line Monitoring

The introduction of new and more reliable sensors and on-line monitors for ammonia, COD, respiration rate, turbidity, and specific pollutants, allows treatment processes to be controlled in real time. This allows a greater control of effluent quality and improves process economics, by allowing the treatment plant to be operated closer to the margins and reduces the need to design in some overcapacity.[21]

Radiation Pulse Treatment

Use of high voltage pulses have been reported to give disinfection, destruction of soluble organics, and rapid dewatering of sludges. The combination of shock wave and soft X-rays could offer significant advantages over conventional treatment, although public acceptance of the technique and its costs have yet to be determined.

Membrane Separation Systems

Recently a number of sewage treatment systems have been developed utilizing membrane filtration. Originally developed for food and pharmaceutical industries, membrane systems have now seen widespread large-scale application in desalination and water treatment plants. This technology relies on a microporous barrier (normally polymeric) in order to filter effluent on the basis of particle size (usually to <1 μm). To reduce fouling, membranes are generally used in a cross-flow arrangement, with the bulk of the effluent flow across the membrane surface through which treated permeate is removed.[22] The process has the advantage of being able to deliver disinfected effluent of uniformly high quality, but can have a significant cost disadvantage compared with conventional treatment. Nonetheless, as perhaps the only truly new sewage treatment process to have been developed in the last ten years, membrane systems have attracted considerable interest from water companies.

Two specific developments are worthy of mention: (i) use of membrane separation as an alternative to biological secondary treatment, followed by disinfection for coastal sites, and (ii) membrane bioreactor systems.

Membrane Secondary Treatment. Screening and primary treatment is followed

[21] G. Ladiges and R. Kayser, *Water Sci. Tech.*, 1993, **28**, 11/12, p. 315.
[22] J. Murkes and C. G. Carlsson, 'Crossflow Filtration—Theory and Practice', John Wiley and Sons, Chichester, 1988.

by membrane separation to remove remaining suspended solids and achieve disinfection by physical removal of pathogenic organisms, including bacteria and viruses. Since discharge of soluble BOD is rarely an issue at coastal sites, membrane treatment can significantly reduce capital expenditure as biological treatment, settlement, disinfection, and outfall costs are decreased. However, operating costs are higher and overall cost effectiveness depends on site-specific factors.[23] Of the systems available, those from Memcor[24] and Renovexx[25] are now seeing full scale application.

Membrane Bioreactor Systems. Membrane bioreactors combine biological treatment and barrier separation stages in one treatment system. Use of membranes to replace settlement tanks allows the retention of high levels of biomass in the biological treatment stage (typically 15 to 20 g l^{-1} mixed liquor suspended solids). This provides a compact system, giving a very high quality disinfected effluent and lower waste sludge volumes due to thicker sludges than conventional proccesses. However, the current cost of membranes limits cost competitiveness to smaller scale systems where high effluent quality or restricted land area is a concern. The fully automated system developed by Kubota, in Japan, probably represents the forefront in advanced sewage treatment systems.[26]

6 Conclusions

The effective and efficient treatment of sewage and industrial effluent is essential to prevent damage to the environment. This has been recognized and enforced by both UK and European legislation. A wide range of solutions will be needed to ensure that the tighter regulatory requirements are achieved on a consistent and more economic basis. In the industrialized nations, improvements to existing assets and optimization of treatment processes will dominate developments in treatment over the next decade.

In the developing world there is a clear need for appropriate low-cost sewage treatment. In these countries the main future requirements will be for sewerage in cities, and the provision of preliminary, primary, and eventually secondary treatment. It is a sad reflection that, over a century on from the discovery linking disease to contaminated water, some 25 000 people die every day as a result of lack of clean water and basic sanitation. Most of these are children in the developing countries. It is estimated that two billion people, nearly half the world population, do not have clean drinking water.

[23] G. Owen, M. Bandi, J. A. Howell, and S. J. Churchouse, 'Economic Assessment of membrane Processes for Water and Wastewater Applications', Proceedings of the 'Engineering of Membrane Processes II—Environmental Applications' Conference, Elsevier, Amsterdam, 1994.
[24] F. Hudman, P. MacInante, A. Day, and W. Johnson, 'Demonstration of Memtec Microfiltration for Disinfection of Secondary Treated Sewage', Sydney Water Board, Memtec Ltd. and Department of Industry, Technology and Commerce, Vol. 1, May 1992.
[25] G. J. Realey and J. Bryan, 'Preliminary Evaluation of the Renovexx Microfiltration System at Berwick upon Tweed STW', WRC Report UM1379, May 1993.
[26] H. Ishida, Y. Yamada, M. Tsuboi, and S. Matsumura, 'Submerged Membrane Activated Sludge Process (KSMASP)—Its Application into Activated Sludge Process with High Concentration of MLSS', Proceedings of the 2nd International Conference on 'Advances in Water Effluent Treatment', MEP, BHR group publication, London, 1993, **8**, p. 321.

Sewage and Industrial Effluents

The charity 'WaterAid', which is supported by the UK Water Companies, is helping to address this problem. In 1993 it provided £6M to the developing countries of India, Nepal, Kenya, Ethiopia, Uganda, Tanzania, Zimbabwe, Ghana, Sierra Leone, and The Gambia, for water supply and basic sanitation. This vital work continues.

Acknowledgement

The authors would like to thank all the Wessex Water staff who assisted in the formulation of this article, *viz.*: Sam Allen, Fiona Bowles, Phil Charrett, Steve Churchouse, Simon Cole, Rikk Earthy, Keith Fitzgerald, Emma Letts, Andrew Randle, Mike Tarbox, and Peter Wratten.

The views expressed in this paper are those of the authors and do not necessarily represent those of Wessex Water.

Landfill

K. WESTLAKE

1 Introduction

The disposal of wastes to land has been the prime means of waste disposal since the evolution of man. Since the late nineteenth century, the volume and hazardous nature of wastes generated has increased considerably, and has led to the need for disposal to land specifically allocated for the purposes of disposal—landfill. Even today, the disposal to land is often poorly controlled and managed, especially in developing countries. This article will not focus on these 'dumps', but on properly managed and controlled landfills. Nor is there scope here for debating in detail the relative merits of landfill disposal and other waste management options and their role in integrated waste management, although these concepts will be introduced.

European Union (EU) policy on waste management is clearly enunciated in the Fifth Environment Action Programme 'Towards Sustainability' [COM(92)23][1] issued in March 1992. The programme sets long-term policy objectives and intermediate targets for the year 2000. For 'Municipal waste' the overall target is the 'rational and sustainable use of resources', achieved through a hierarchy of management options, *viz.*

- Prevention of waste.
- Recycling and re-use.
- Safe disposal of remaining waste in the following rank order:
    (i) combustion as fuel;
    (ii) incineration;
    (iii) landfill.

Thus, it is clear that the European Union views landfill as the final waste disposal option. This view has been translated into potentially increasingly stringent controls over landfill as identified within the draft Council Directive on the landfill of waste [COM(93)275].[2] The potential impact and the requirements of

---

[1] Commission of the European Communities, 'Towards Sustainability: A European Programme of Policy and Action in Relation to the Environment and Sustainable Development', COM(92)23, 1992.
[2] Commission of the European Communities, 'Proposal for a Council Directive on the Landfill of Waste', COM(93)275, 1993.

the Directive on the landfill wastes within Europe are too lengthy to discuss in detail here. However, at the time of writing, some of the main requirements are that

- clinical wastes will be banned from landfill;
- leachate and groundwater will be monitored at least twice per year for at least 30 years from the date of implementation;
- leachate collection/drainage systems should be sufficient to ensure that no liquid accumulates at the bottom of the site;
- landfill gas will have to be collected and treated unless an environmental assessment determines that this is not required; and
- a conditioning plan identifying the measures to comply with the Directive should be submitted to the competent authority within one year of the implementation of the Directive.

At the time of writing, the future of co-disposal (see Section 3) remains uncertain. Latest reports[3,4] suggest that co-disposal will be allowed in countries where currently practised subject to conditions which are to be met within five years of implementation of the Directive. Existing landfills, or those created between 1994 and the adoption of the Directive, will have to meet the required standards within ten years of adoption. However, the wording of the revised document appears to allow for different interpretations, and uncertainty still remains.

Requirements such as those above represent a significant step forward for a number of countries within the EU; according to a recent report by Environmental Resources Management (cited in Reference 5), the number of uncontrolled landfills in Italy and Portugal exceeds 60% of the total number of sites, while in Greece and Spain the number of uncontrolled sites represent approximately 30% of the total. Strict implementation of the Landfill Directive could result in a decrease in landfill capacity in Europe as sites opt for closure rather than continued operation under new specifications that for many would be both difficult and expensive to achieve.

In those countries whose landfill operations are better controlled, the impact of the Directive will still be significant; in the United Kingdom (UK) there are approximately 4000 licensed landfill sites where approximately 85% of controlled wastes and 70% of hazardous wastes (equivalent to approximately 2 million tonnes annum^{-1}) are disposed. The proposed ban on co-disposal will, according to the Department of Environment (DoE),[6] add an extra £160M year^{-1} to UK industries' waste disposal costs. As a result of engineering and other requirements of the Directive, the cost of landfill disposal can also be anticipated to increase significantly. As the cost increases, and the differential between landfill and other disposal options such as incineration decreases, so the easier it becomes to use alternative disposal routes that are more favourably placed in the waste treatment hierarchy. In this way, the objectives of the EU Fifth Environment Action Programme begin to become achieved. Also, although cost is obviously

[3] S. Tromans, *Wastes Manage.*, 1994, July, 16.
[4] *Ends Rep.*, 1994, **233**, 34.
[5] *Warmer Bull.*, 1994 **41**, 2.
[6] *Ends Rep.*, 1994, **228**, 38.

Landfill

an important factor, in countries such as the Netherlands, Denmark, and Japan, where local geology cannot support landfill as easily as in the UK, there is also a greater political will to find alternatives to landfill.

This review will examine the science, engineering and control of landfill disposal in the light of the above influences, using UK practice and control methods to highlight changing trends in landfill policy and practice.

2 Principles of Landfill Practice

Landfill has been defined[7] as 'the engineered deposit of waste onto and into land in such a way that pollution or harm to the environment is prevented and, through restoration, land provided which may be used for another purpose'. Prevention of 'harm to the environment' is achieved in a number of ways, but requires effective control of waste degradation processes and effective landfill design, engineering, and management. The principles of landfill practice used to achieve this have changed considerably since the 1970s and three major principles of landfill design and management have been recognized; these are 'dilute and attenuate' (otherwise known as 'dilute and disperse'), 'containment', and 'entombment'.

Dilute and Attenuate

Dilute and attenuate is the principle of landfill disposal for unconfined sites, with little or no engineering of the site boundary, in which leachate [that liquid formed within a landfill site that is comprised of the liquids that enter the site (including rainwater) and the material that is leached from the wastes as the infiltrating liquids percolate downwards through the waste] formed within the waste is allowed to migrate into the surrounding environment. This principle relies upon attenuation of the leachate both within the waste and in the surrounding geology, by biological and physico-chemical processes. Dilution within groundwater further reduces the risk posed by the migrating leachate—but, by definition, necessarily contaminates that groundwater. For the 'dilute and attenuate' principle to be effective, the associated risk should be deemed to be acceptable.

Until the 1980s most landfills were based on the 'dilute and attenuate' philosophy, the principle being supported by studies and in the mid 1970s;[8] these showed attenuation of leachate as it moved through various unsaturated strata, and that the attenuation processes (defined to include dilution) could be used to effectively treat landfill leachate as it migrated from the site. The advantages of the 'dilute and attenuate' landfill are that there is no requirement for expensive landfill lining/engineering, and as liquids formed within the site migrate from the base, there is no requirement for leachate collection and treatment facilities. Unlike countries such as the Netherlands where the water-table is very close to the surface, groundwater within the UK is often located in deep aquifers,

[7] J. Skitt, '1000 Terms in Solid Waste Management', ISWA, Denmark, 1992, p. 93.
[8] Department of the Environment, 'Co-operative Programme of Research on the Behaviour of Hazardous Wastes in Landfill Sites. Final Report of the Policy Review Committee', HMSO, London, 1978.

facilitating attenuation in the unsaturated strata above. However, such attenuation cannot be relied upon in all circumstances and, although groundwater and surface waters appeared free of contamination by landfill leachate for many years, recent examples[9,10] have shown that some contamination has occurred. That this has happened is due to the fact that 'dilute and attenuate' was adopted irrespective of local conditions and without prior risk assessment. A large number of sites that were developed prior to the 1990s, both within the UK and elsewhere, continue to operate on the 'dilute and attenuate' principle. For some sites with appropriate geology/hydrogeology, and with a suitable, supporting risk assessment, landfills based on the 'dilute and attenuate' principle may still be technically feasible. However, for political and other reasons, any such landfills are now unlikely to be built in the developed world, irrespective of local conditions.

In 1980, the introduction of the European 'Groundwater Directive' (The Protection of Groundwater against Pollution caused by Certain Dangerous Substances—80/68/EEC)[11] caused a reassessment of the 'dilute and attenuate' principle. The Groundwater Directive prohibited the direct or indirect discharge into groundwater of List I (most potentially polluting) substances and limited discharges of List II substances unless prior investigation showed that pollution of groundwater would not occur, or unless the groundwater was permanently unsuitable for other purposes. Because many of the substances present in List I and List 2 (Table 1) could be found in landfill leachate, it was clear that better control would have to be exercised. However, the Groundwater Directive did not address issues of diffuse pollution nor the management of abstraction or groundwater monitoring. Recognizing this, and in response to duties under the Water Resources Act (1991), the National Rivers Authority (NRA) within the UK, issued a groundwater protection policy[12] in which NRA policy towards the protection of groundwater was identified. The introduction of this policy within the UK is one of a number of factors that has encouraged a move towards the containment landfill.

Containment

The containment principle of landfill requires a much greater degree of site design, engineering, and management, and exercises some degree of control over the hazards associated with the disposal of wastes to landfill. In the developed world, containment landfill is now the accepted means of disposal to land, although the degree of engineering to achieve containment, and the management of water and other parameters varies considerably.

The underlying principle of containment landfill is that liquids (leachate) generated within the waste should not be allowed to migrate beyond the site boundary. A 'Containment site' has been defined[13] as a 'landfill site where the

---

[9] *Ends Rep.*, 1993, **225**, 6.
[10] *Ends Rep.*, 1994, **229**, 12.
[11] Commission of the European Communities, 'The Protection of Groundwater Against Pollution caused by Certain Dangerous Substances', 80/68/EEC, 1980.
[12] National Rivers Authority, 'Policy and Practise for the Protection of Groundwater', National Rivers Authority, 1992.
[13] J. Skitt, '1000 Terms in Solid Waste Management', ISWA, Denmark, 1992, p. 49.

Table 1 List I and List II substances as defined by the EC Groundwater Directive (80/68/EEC)

List I of Families and Groups of Substances	*List II of Families and Groups of Substances*[a]
List I contains the individual substances which belong to the families and groups of substances specified below, with the exception of those which are considered inappropriate to List I on the basis of a low risk toxicity, persistence and bioaccumulation. Such substances which with regard to toxicity, persistence, and bioaccumulation are appropriate to List II are to be classed in List II. (1) Organohalogen compounds and substances which may form such compounds in the aquatic environment. (2) Organophosphorus compounds. (3) Organotin compounds. (4) Substances which possess carcinogenic, mutagenic, or teratogenic properties in or via the aquatic environment (1). (5) Mercury and its compounds. (6) Cadmium and its compounds. (7) Mineral oils and hydrocarbons. (8) Cyanides.	List II contains the individual substances and the categories of substances belonging to the families and groups of substances listed below which could have a harmful effect on groundwater. (1) The following metalloids and metals and their compounds: Zinc Copper Nickel Chrome Lead Selenium Arsenic Antimony Molybdenum Titanium Tin Barium Beryllium Boron Uranium Vanadium Cobalt Thallium Tellurium Silver (2) Biocides and their derivatives not appearing in List I. (3) Substances which have a deleterious effect on the taste and/or odour of groundwater and compounds liable to cause the formation of such substances in such water and to render it unfit for human consumption. (4) Toxic or persistent organic compounds of silicon and substances which may cause the formation of such compounds in water, excluding those which are biologically harmless or are rapidly converted in water into harmless substances. (5) Inorganic compounds of phosphorus and elemental phosphorus. (6) Fluorides. (7) Ammonia and nitrates.

[a] Where certain substances in List II are carcinogenic, mutagenic, or teratogenic they are included in category 4 of List I.

rate of release of leachate into the environment is extremely low. Polluting components in wastes are retained within such landfills for sufficient time to allow biodegradation and attenuating processes to occur, thus preventing the escape of polluting species at an unacceptable concentration'. According to the site engineering, gas migration may also be prevented or significantly reduced. Perhaps the key phrases in the above definition are 'rate of release ... is extremely low' and 'at an unacceptable concentration', for this implies that some migration of leachate may occur, but the associated risk is acceptable. Even the most highly engineered containment landfills must be expected to fail at some time in the future, whereupon leachate will be released. Recognizing this, the operation and management of landfills should be undertaken in such a way that any release will be at an **acceptable** concentration. The application of risk assessment to containment landfill has been discussed elsewhere.[14] With the concept of acceptable and managed risk in mind, the concept of the sustainable landfill,[15] and fail-safe landfill[16] have been proposed.

The containment of leachate implies, in most cases (see Entombment), that liquid (leachate) will collect and will require treatment. This has placed new requirements upon the effective management of landfills (see Leachate Management).

Entombment

Entombment landfill is based upon the principle of containment landfill, but attempts to prevent infiltration of liquids, thereby storing the waste, in perpetuity, in a relatively dry form. Waste storage in this way accepts that no attenuation of wastes will occur and argues that storage creates the opportunity for development of new technologies to deal with the stored wastes in a more appropriate way at some time in the future. The converse argument is that waste treatment, not storage, is the only option available under the concept of sustainable development. 'Sustainable development' was defined in 1987 by the World Commission on Environment and Development[17] as 'development that meets the needs of the present without compromising the ability of future generations to meet their own needs'. The concept of the sustainable landfill as an appropriate method of waste treatment has been proposed recently.[15]

Sustainable Landfill

For waste management, the sustainable landfill could be crudely interpreted as dealing with today's waste today and not passing on today's waste for the future generations to deal with. The proposed strategy accommodates both the landfill

[14] J.I. Petts, 'Proceedings 1993 Harwell Waste Management Symposium: Options for Landfill Containment', AEA Technology, Harwell, 1993.
[15] R.C. Harris, K. Knox, and N. Walker, *IWM Proc.*, 1994, Jan, 26.
[16] M. Loxham, 'Proceedings Landfill Tomorrow—Bioreactor or storage', Imperial College of Science, Technology, and Medicine, London, 1993.
[17] World Commission on Environment and Development, 'Our Common Future', Oxford University Press, Oxford, 1987.

Landfill

disposal of untreated waste and the achievement of final storage quality within 30 years. For this to occur, waste must either be pre-treated to a state close to final storage quality or stabilization within the landfill must be accelerated. However, while the theory is sound, so far little has been done at the scale of an operating landfill to demonstrate the effectiveness of enhancement techniques for waste degradation. Whether because of this uncertainty, or because of more political factors, a number of European countries appear to be opting for pre-treatment as the means of achieving sustainable landfill disposal. Germany, for example,[18] has limited the amount of organic waste disposed to landfill to 5%, although where this is not feasible the relevant bodies have given until 2005 to achieve this goal.

Fail-safe Landfill

The concept of the 'sustainable landfill' is echoed in the philosophy of fail-safe landfill.[16] The fail-safe philosophy argues that whatever the containment system utilized, it will ultimately fail and/or institutional control will cease, and the contents within, *e.g.* leachate, will be released to the environment. It therefore requires that any releases should be such that the risk posed to the environment is acceptable. For this to be the case, wastes disposed to landfill must again be pre-treated or degradation must be accelerated such that the hazardous nature of the wastes and waste products are minimized.

The time taken to liner failure will be dependent upon a number of factors, some of which may be sudden, catastrophic, and unpredictable. Others such as landfill engineering, design, and control are considered later. Thirty years is often considered as a suitable lifetime expectancy for synthetic liners, although there are many examples of liner failure within this period.

In order to minimize the risks associated with, amongst other things, liner failure, landraising has been considered as an alternative to landfill.

Landraising

Landraising (defined here as the emplacement of waste with the base at ground level rather than within a hole), is currently receiving greater support within scientific and technical circles, although within the UK, a number of planning applications for landraising have been turned down recently on the grounds of 'loss of amenity'.[19] Typical concerns relate to visual impact, noise, and landscaping, and for landraising to be effective, such concerns must be balanced against the potential environmental benefits. Landraising, because of the facility for increased control over emissions (*e.g.* collection of leachate and subsequent treatment is much more easily achieved, while the opportunity for gas migration through geological strata is considerably reduced) may also allow landfill development in areas otherwise considered to be too vulnerable for landfill development. Vulnerable sites include those located above aquifers, while those sited above clay, for example, could be considered as being suitable for development either with or without a complementary lining system.

[18] K. Stief, 'Proceedings of the Fourth International Landfill Symposium', Cagliari, 1993.
[19] *Waste Plan.*, 1994, **11**, 20.

Table 2 Comparison of waste characteristics from different countries (as % waste arisings)

	United Kingdom	Asian city	Middle East city
Vegetable	28	75	50
Paper	37	2	16
Metals	9	0.1	5
Glass	9	0.2	2
Textiles	3	3	3
Plastic	2	1	1
Miscellaneous	12	18.7	23
Weight person^{-1} day^{-1}	0.845 kg	0.415 kg	1.060 kg
Density kg m^{-3}	132	570	211

Source: J. R. Holmes.[60]

3 Landfill Processes

Waste Composition

The typical composition of wastes disposed to landfill within the UK, an Asian city, and a Middle East city is shown in Table 2. From the data presented, it is clear that both the waste composition and the rate of waste generation vary considerably from country to country. These variations will have a marked impact on both the processes that occur within landfill and upon the management strategies required to effectively control the landfill development. According to Holmes[20] 'Quantities of waste are invariably lower in developing countries because of lower prosperity and consumption as well as extensive scavenging by beggars and the very poor. Densities of waste are much higher because of the absence of paper, plastics, glass, and packaging materials and hence a much greater concentration of putrescible matter. Moisture contents at 40–50% are much higher than those in developed countries at 20–30%'. In developing countries, landfill is often poorly controlled and presents a unique set of problems, beyond the scope of this article. Waste management practice in developing countries has been summarized elsewhere.[20]

Even within any one country, waste composition can vary considerably over time and from season to season. For example, within the UK the composition of Municipal Solid Waste (MSW) changed considerably after the introduction of the Clean Air Act (1956). This Act resulted in a reduction of organic material (such as newspaper and vegetables) burned in the fireplace, and a concomitant increase in the same material disposed to landfill—with important ramifications to the potential for landfill gas (see Section 6) and landfill leachate (see Section 5) production.

Waste Degradation

Although many of the processes thought to occur within landfill have not been proven, the presence of predicted intermediate products and end products of

[20] J. R. Holmes, *Wastes Manage.*, 1992, June, 8.

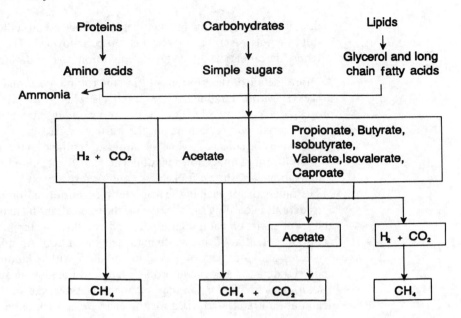

Figure 1 Landfill waste degradation processes. Source: Waste Management Paper 26[50]

degradation, together with the presence of relevant enzymes, leads us to conclude that the degradation of organic wastes in the landfill environment is similar to the degradation of organic materials in other anaerobic environments.

When deposited within the landfill, oxygen entrapped within the void spaces is rapidly depleted as a result of biological activity, and the local environment becomes anaerobic, encouraging the growth of anaerobic micro-organisms, especially bacteria. Carbon dioxide and methane are produced as a result of anaerobic microbial activity, and displace nitrogen remaining from the entrapped air. Eventually a dynamic equilibrium is reached with a gas ratio within the landfill of approximately 60 methane:40 carbon dioxide.

Organic materials, such as paper and vegetables, which are essentially comprised of carbohydrate, protein, and lipid will breakdown as shown in Figure 1. In the initial stages of degradation, the polymeric compounds such as cellulose will be hydrolysed to their component parts. This reaction is effected by bacteria within the waste materials. The sugars produced are used by bacteria for carbon and energy metabolism, producing in turn a range of organic acids which serve as substrate for the growth of methanogenic bacteria, producing methane as an end-product. The fate of inorganic materials will vary according to the compound/element of interest. Some of particular interest are discussed briefly below:

Sulfate. Sulfate-containing material such as plaster board (where the sulfate is present as gypsum) will be reduced under anaerobic conditions and will associate with free metal ions to produce a metal sulfide. Under acidic conditions the sulfate will be released as hydrogen sulfide and as such can present a significant hazard.

Heavy metals. Reduced sulfate, as sulfide, will react with heavy metals, as described above, producing relatively immobile metal sulfide. Other insoluble salts such as metal carbonates may also be formed. The overall effect is to reduce the potential for heavy metal migration beyond the landfill site boundary.

A more detailed description of the microbiological and physico-chemical processes within landfill can be found elsewhere.[21-23] As a result of these processes, a 'cocktail' of solids, liquid, and gaseous chemicals is produced. The solid materials will remain within the waste, the gases may migrate beyond the waste or may be collected, while the liquid and soluble components will, together with infiltrating liquids, form landfill leachate. Typical landfill gas and leachate compositions are shown in Tables 3 and 4, respectively.

The timescale for each of the degradation/transformation processes may vary considerably according to the nature of the wastes, landfill management practice, and local environmental conditions, and can have a significant impact on the pollution potential of this waste management option. An understanding of the processes and products of waste degradation within landfill is important to understanding the associated problems, and to the identification of appropriate resolutions. An understanding of the processes that occur during waste degradation are also required for the effective control of co-disposal.

Co-disposal Landfill

As discussed earlier, the future of co-disposal as a waste treatment option within Europe remains uncertain. Co-disposal is[24] 'the calculated and monitored treatment of industrial and commercial, liquid and solid wastes by interaction with biodegradable wastes in a controlled landfill site'. The philosophy behind co-disposal is that the microbiological and physico-chemical processes that occur during the degradation of organic wastes (such as those contained within municipal solid wastes) will **treat** the co-disposed wastes and reduce the associated hazards. Thus the landfill is regarded as a biological reactor where treatment, rather than storage, occurs.

The co-disposal of industrial and commercial wastes has been common practice in the UK for many years and is carried out in many other countries, although elsewhere it may not be called co-disposal. While the waste industry within the UK generally favours co-disposal as a means of waste treatment, it has been unable to provide sufficient evidence in support of the argument that co-disposal is safe, and thus it has proved difficult to dispute counter arguments that co-disposal represents unacceptable practice.

The removal of co-disposal as a waste treatment option would have significant ramifications for waste management across Europe. In the UK, for example, where approximately 70% of a total of 2.8M tonnes of difficult industrial wastes are co-disposed to landfill,[25] alternative treatment would be required for

[21] E. Senior, 'Microbiology of Landfill Sites', CRC Press, Boca Raton, Florida, 1990.
[22] M. A. Barlaz, R. K. Ham, and D. M. Schaefer, *Crit. Rev. Environ. Control*, 1990, **6**, 557.
[23] K. Westlake, 'Proceedings International Conference on Landfill Gas: Energy and Environment', Bournemouth, 1990.
[24] J. Skitt, '1000 Terms in Solid Waste Management', ISWA, Denmark, 1992, p. 93.
[25] *Ends Rep.*, 1994, **228**, 38.

Table 3 Typical landfill gas composition (% vol.)[1]

Component	Typical value (mature refuse)	Observed maximum	Reasons for component being unusually abundant
Methane	63.8[2]	77.1	Adsorption of carbon dioxide (*e.g.* by water, lime)
Carbon dioxide	33.6[2]	89.3	Aerobic degradation of refuse
Oxygen	0.16[4]	20.9[4]	Air mixed with landfill gas
Nitrogen	2.4[2]	80.3	Air mixed with landfill gas, or very slow degradation if oxygen depleted
Hydrogen	<0.05	21.1	Young refuse. Methane concentration usually low
Carbon monoxide	<0.001	—[5]	Oxygen-starved burning in refuse
Saturated hydrocarbons	0.005[3]	0.074	Young refuse or high concentrations of petrochemicals present
Unsaturated hydrocarbons	0.009[3]	0.048	Young refuse or petrochemicals/solvents present
Halogenated compounds	0.00002[3]	0.032	Young refuse or solvents present
Hydrogen sulfide	0.00002[3]	0.0014 35[6]	Young refuse High sulfate waste present
Organosulfur compounds	<0.00001[3]	0.028	Young refuse
Alcohols	<0.00001[3]	0.127	Young or semi-aerobic refuse (fermentation)
Others (not included above)	0.00005[3]	0.023	Solvents or other volatile wastes deposited

[1] Landfill gas usually emerges saturated with water vapour, representing 0.001% to 0.004%, depending on its temperature.
[2] Based on long term data from Stewartby landfill, supplied by London Brick Landfill Limited.
[3] Based on five year old refuse.
[4] Entirely derived from atmospheric oxygen.
[5] Concentrations of several per cent carbon monoxide has been reported at landfills on fire but have not been confirmed.
[6] Refuse mixed with plasterboard.

approximately 2M tonnes of such wastes.

According to Knox and Gronow,[26] for effective co-disposal it is important that only those wastes that can be effectively treated are selected, and that

- only containment sites are utilized;
- the process is effectively managed and controlled;
- co-disposal occurs into biologically active waste at rates not exceeding recommended loading rates; and
- effective monitoring and after-care including effective monitoring of both gas and leachate are undertaken.

[26] K. Knox and J. Gronow, *Waste Manage. Res.*, 1990, **8**, 255.

Table 4 Typical composition of leachates from domestic wastes at various stages of decomposition (all figures in mg l^{-1} except pH values)

Determinands	Fresh wastes	Aged wastes	Wastes with high moisture contents
pH	6.2	7.5	8.0
COD	23 800	1160	1500
BOD	11 900	260	500
TOC	8000	465	450
Volatile acids (as C)	5688	5	12
NH_3-N	790	370	1000
NO_3-N	3	1	1.0
Ortho-P	0.73	1.4	1.0
Cl	1315	2080	1390
Na	9601	300	1900
Mg	252	185	186
K	780	590	570
Ca	1820	250	158
Mn	27	2.1	0.05
Fe	540	23	2.0
Cu	0.12	0.03	—
Zn	21.5	0.4	0.5
Pb	0.40	0.14	—

Source: Waste Management Paper 26A.[41]

Wastes that have been identified as being suitable for co-disposal include brewery wastes, animal and food industry wastes, aqueous organics, paint waste, acids, and alkalis. Wastes that are not suitable for co-disposal include flammable wastes, wastes containing PCBs and similar compounds, and acid tars.[26]

The ability of decomposing waste to attenuate added organic and inorganic material has been recognized for many years and a large number of articles have been published, including a major review of co-disposal practice in the UK.[27] For effective co-disposal, the material to which co-disposed wastes are added must be (micro)biologically active, the most effective measure of microbial activity being the production of methane. Methanogenic waste indicates that the biological processes are relatively stable, and that pH is controlled around neutrality.

According to Knox and Gronow,[26] methanogenic waste provides an aqueous chemical environment similar to anaerobic digesters with low redox potential (E_h), near neutral pH, and which is maintained in a buffered steady-state by on-going degradation processes. Those wastes most extensively studied within co-disposal are phenols, cyanides, acids, and heavy metals, where effective degradation and attenuation has been shown. The above review gave the following conclusions on co-disposal.

(i) For phenols, cyanides, heavy metals, and acids, the degradation rates were similar to other types of anaerobic digester.

[27] K. Knox, 'A Review of technical aspects of co-disposal (PECD 7/10/214)', Department of the Environment Report No. CWM 007/89, HMSO, 1989.

Landfill

    (ii) It is likely that the capabilities of methanogenic refuse will extend outside the chemical groups reviewed.
    (iii) The efficiency of the waste was maximized by
        (a) saturated conditions with leachate recycle,
        (b) established methanogenesis, and
        (c) elevated temperatures.

With regard to full-scale co-disposal sited in the UK, the review findings were as follows.

    (i) The major examples of UK co-disposal sites do act as reactors—a wide range of difficult wastes were converted to a low hazard effluent.
    (ii) Organic loading rates were in the range 1–10 g TOC m^3 day^{-1}.
    (iii) Where calculated, heavy metal loadings were of a similar concentration to background levels.

The above studies appear to indicate that effective co-disposal of selected waste streams can be achieved. However, effective management and control is imperative, and while landfill sites retain licences to dispose of wastes not thought suitable for co-disposal, there is likely to be concern about the potential for harm to the environment caused by this practice.

While such concern may (in the case of bad management and control) be justified, alternatives such as mono-disposal or waste storage appear to be potentially more hazardous.

Mono-disposal

If co-disposal is phased out, then mono-disposal represents an alternative option. Mono-disposal (mono-landfill or monofill) requires that wastes that are similar in nature are emplaced together, but that mixing with other wastes should not occur.

When undertaking mono-disposal, the buffering capacity and pH control which are typical of co-disposal sites are likely to be lost, and biological and physico-chemical processes of degradation and attenuation are unlikely to occur. Thus, the emplaced wastes are likely to remain relatively unchanged and the associated risk to the environment will also remain unchanged. In this way, mono-disposal is similar to waste storage and as such is an anathema to the concept of sustainable development, although for some wastes there may be little associated risk, and in this case mono-disposal may be appropriate. For example, it could be argued that the mono-disposal of asbestos wastes has a low associated risk and as such would be an appropriate option. As with any other form of landfill, failure of the containment system must be anticipated. At such time, the associated risk will, because there will have been no, or limited, opportunity for attenuation, be much higher than had the same wastes been co-disposed. Arguments that mono-disposal facilitates development of technologies in the future to deal more effectively with the stored wastes are again out of step with sustainable development.

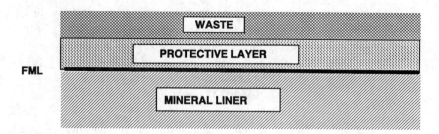

Figure 2 'Composite' landfill liner (FML = flexible membrane liner)

4 Landfill Design, Engineering, and Control

Most modern landfills are now designed as containment landfills and therefore the major design considerations relate to the design of the containment system; this may vary according to local and national policy and according to the landfill location. Within the UK, and NRA's Groundwater Protection policy stressed the importance to groundwater of pollution, and the NRA is currently in the process of producing groundwater vulnerability maps. These maps are based on information on the geological strata (*e.g.* lithological type, permeability characteristics) and the physical properties of the overlying soils such as leaching characteristics. These maps, together with information such as that relating to the location of major abstraction sources, may be used, in the future, to help in the selection of suitable sites for landfill, precluding development on sites of high vulnerability (*e.g.* close to major abstraction source or sited on 'vulnerable' aquifer). This may in turn lead to a more strategic approach to landfill site selection where, for a particular region, the area of least vulnerability would be identified for landfill development. This approach would be quite different to the current market-led strategy.

Landfill Liners

The design and engineering of landfill liners has received much attention in recent years;[28–30] two fundamental types of lining material are available—natural (*e.g.* clay, shale) and synthetic liners, also known as flexible membrane liners (FMLs) or geomembranes. Combination of the two types allow the construction of composite and multiple liners (Figure 2).

Natural liners such as clay have the advantage of inherent attenuation capacity (a relatively high ion-exchange capacity will inhibit, for example, the migration of heavy metals); they are relatively stable in the presence of a wide range of organic and inorganic compounds but they are more permeable than FMLs.[29] Conversely, FMLs[30] have little or no inherent attenuation capacity, are sensitive to organic solvents, but are relatively impermeable. The complementary properties of

---

[28] North West Waste Disposal Officers, Leachate Management Report, Lancashire County Council, 1991.

[29] K. J. Seymour and A. J. Peacock, 'Proceedings of the Second International Landfill Symposium', Sardinia, 1989.

[30] P. J. McKendry, 'Proceedings of the Fourth International Landfill Symposium', Sardinia, 1993.

Table 5 Calculated flow rates (measured in l ha^{-1} day^{-1}) through landfill liners

Type of liner	Best case	Average case	Worst case
Geomembrane alone	2500 (2 holes ha^{-1})	25 000 (20 holes ha^{-1})	75 000 (60 holes ha^{-1})
Compacted soil alone	115 ($k = 10^{-10}$ m s^{-1})	1150 ($k = 10^{-10}$ m s^{-1})	11 500 ($k = 10^{-10}$ m s^{-1})
Composite	0.8 (2 holes ha^{-1}, $k = 10^{-1}$ m s^{-1} Poor contact)	47 (20 holes ha^{-1}, $k = 10^{-9}$ m s^{-1} Poor contact)	770 (60 holes ha^{-1}, $k = 10^{-8}$ m s^{-1} Poor contact)

k = coefficient of permeability
Source: A. Street[61]

natural and synthetic materials are optimized in the construction of composite liner systems where the FML is placed in intimate contact above a natural liner. For much of the developed world, the simple composite liner is considered as the minimum requirement, while in the USA, and increasingly within Europe, multiple liner systems in which multiple barriers with protective layers, and monitoring layers, are preferred.

However, multiple-layer systems do not necessarily provide enhanced environmental protection. After loading with waste, multiple-liner systems have been known to slip due to sheer forces, with resultant failure of containment. There is also considerable debate concerning the relative merits of monitoring or drainage layers; it can be argued that this facility allows the recognition of containment failure and remedial action to be taken before serious pollution occurs. Conversely, the creation of a permeable layer across the whole base of a site effectively converts a point source leak to one that may cover many hectares with significant increase in the associated risk, and effective weakening of the containment system as a whole. Table 5 presents relative leakage rates from different lining systems, and demonstrates the reduction in leakage rate that can be achieved by using a simple composite liner rather than either a natural liner alone or an FML alone.

Leachate Management

There is a wide variety of leachate treatment systems available, including the aerated lagoon, the rotating biological contractor, air stripping, and reed beds. Such systems have been reviewed elsewhere.[31,32]

The continuing debate in leachate management is whether or not to add water to sites or to allow water infiltration. In the USA there has been a requirement since October 1993 to prevent rainfall infiltration into household waste under the Resource Conservation and Recovery Act (RCRA) subtitled 'D'.[33] This approach has been called the 'dry tomb' approach as it requires the design and management of landfills in such a way as to minimize liquid infiltration into the

---

[31] H. D. Robinson, *J. Inst. Water Environ. Manage.*, 1990, **4**, 78.
[32] H. D. Robinson, M. J. Barr, and S. D. Last, *J. Inst. Water Environ. Manage.*, 1992, **3**, 321.
[33] US EPA Solid Waste Disposal Facility Criteria; Final Rule 40 CFR Parts 257 and 258, *Fed. Reg.*, 1991, **56**, 50978.

waste. This principle is contrary to that engendered by the 'sustainable landfill' and fail safe landfill, in that with the dry tomb approach a final storage quality waste will not be produced, and there will always remain future pollution potential. While it may be possible to ensure effective containment and capping in the short-term (measured in tens of years), ultimate long-term failure (measured in tens or hundreds of years) of lining and capping systems must be anticipated. The arguments for sustainable landfill design with moisture control are considerably stronger.

The requirements for effective control of landfill leachate are becoming increasingly more stringent and are considered in more detail in Section 7. For example, the recommended maximum value for ammonia in landfill leachate is 5 mg l^{-1}. Within the landfill environment, there are no recognized pathways for the complete removal of nitrogen (as ammonia) and calculations based on flushing rates alone have shown that up to 500 years or more would be required to achieve completion criteria levels.[34] Clearly this does not meet the requirements of sustainable development and considerable management and treatment of leachate will be required.

However, recent research[35] has shown that nitrification of ammonia within landfill leachate can be achieved in aerobic treatment plants, and that recycling of the nitrate produced, through landfill waste, can then convert the nitrate to nitrogen gas, thus effectively removing the ammonia from leachate. However, field-scale trials will be required before the true effectiveness can be assessed. Whether or not liquid addition is required to increase the hydraulic retention time will, to an extent, be determined by the moisture content of the waste at emplacement. For instance, it is known that the loading of waste with more waste layers will 'squeeze out' liquids absorbed within the lower layers. Recirculation of this leachate may then provide sufficient leachate recycle to flush out nitrogen (as ammonia) within the upper layers.

It is clear from the above that the landfill of the future will require a much more sophisticated leachate management system and that his may require more effective and more rigorous control. Also, engineering and other problems will have to be surmounted; these include the effective collection of leachate and subsequent redistribution (recycle) throughout the waste.

Leachate Drainage and Collection.   For some existing landfill sites there will be no facility for leachate drainage. Some sites will have leachate drains emplaced within 'no-fines' stone and arranged in a variety of patterns (*e.g.* Herring bone) across the site. Many new sites are required to have a drainage blanket across the full base of the site. Typically, the drainage blanket will comprise no fines stone laid at a gradient of 2% to a collection sump. The drainage blanket will usually incorporate an underdrainage system with perforated High Density Polyethylene (HDPE) (or similar) pipework to facilitate drainage to the sump. The spacing between the HDPE drains will vary according to the porosity of the drainage medium and the base gradient, but is often approximately 20 m between adjacent

[34] K. Knox, 'Proceedings International Conference on Landfill Gas: Energy and Environment', Bournemouth, 1990.
[35] K. Knox, 'Proceedings of the Fourth International Landfill Symposium', Sardinia, 1993.

drains. Drain diameters are often 150 mm, although larger diameter drains (300 mm) are becoming more popular due to facilitated inspection and cleaning capabilities. Further consideration of leachate drainage and collection facilities can be found elsewhere.[28]

In recent years there has been an increasing tendency to use geotextiles in the construction of drainage and lining systems. According to the nature of the geotextile, they are used for the provision of extra strength or to provide protection. In drainage systems, they have been used to cover HDPE drains to prevent blockage of the perforations within the HDPE, although there is concern that when used as such, the geotextiles may themselves become blocked by fines or by growth of bacterial film on the geotextile surface.

Sumps and pumps should be reliable, durable, and of sufficient strength to withstand the forces of waste settlement. According to the size of the site there may be a requirement for more than one sump from which the leachate may be pumped. In shallow sites, collected leachate may be withdrawn by vacuum tanker, while in deeper sites (*e.g.* 30 m) the use of vacuum systems is not possible. Here, the use of pumps is required. Ideally, the pipework from the leachate drainage and collection should not penetrate the liner systems, and when planned from the initial stages of landfill design the pipework can be routed to the surface via the side of the landfill. Landfill leachate control has been reviewed elsewhere.[36]

Leachate Recycle. The scientific arguments for leachate recycle, such as reduction in leachate organic and ammonia content, are becoming stronger. One of the major difficulties associated with leachate recycle is that related to the engineering of safe and effective redistribution of leachate throughout the waste. To date, most leachate recycling has been undertaken by means of spraying the leachate across the landfill surface. This method has the advantage that some water may be lost through evaporation, and thus cause a reduction in the volume of leachate for ultimate treatment, although the creation of metal oxide/hydroxide deposits on the surface of the landfill can also prove to be unsightly. The use of under-cap recycle systems provide an alternative to spraying the leachate and will reduce the risks associated with the creation of aerosols that may contain hazardous materials or bacteria. Such a system of perforated HDPE pipes can be placed within a layer of stone placed directly above the waste. The major drawback with this method is that it is difficult to construct a system capable of coping with differential settlement without fracturing.

5 Landfill Gas

Landfill Gas Control

There are a large number of risks associated with the escape of landfill gas, including health risks, explosion risks, and risks associated with atmospheric pollution. Upon closure and capping of a landfill, the primary route of landfill gas migration (via the waste surface) is considerably restricted. Although it has been shown that gas can migrate through clay relatively easily,[37] the landfill cap

---

[36] N. Walker, *Wastes Manage. Proc.*, 1993, Jan, 3.
[37] P. Parsons, 'Proceedings 1993 Harwell Waste Management Symposium: Options for Landfill

nevertheless creates a barrier that restricts movement to the surface with a concomitant increase in gas pressure within the waste. Under these conditions, fractured or porous sub-surface strata may provide the path of least resistance to gas movement. The migration of gas beyond landfill boundaries has been the cause of a number of hazardous (explosion-related) incidents[38] one of the most notable within the UK resulting in destruction of a bungalow at Loscoe in Derbyshire.[39] There is also increasing concern over the contribution of landfill gas to 'global warming'. Recent estimates suggest that landfill emissions of methane could increase more than three-fold over the next 30 years and that landfills are the single largest source of methane in the environment within the UK.[40] Health hazards associated with landfill gas have been discussed elsewhere.[38] These studies[38] on the composition of landfill gas have shown the presence of a large number of toxic trace gases, even in those gases released after flaring, and concerns over the harmful nature has caused the imposition of controls on fugitive gases from the surface of landfills in the USA,[38] where volatile organic compounds and toxic components are of particular concern.

It is clear that effective gas control is an important part of landfill management. The science of gas control has increased considerably in recent years, but relies upon two main mechanisms, *viz.*

(i) use of impermeable barriers, and
(ii) the creation of paths of least resistance through the use of vents, wells, and trenches.

Vents, wells, and trenches that are either actively (using pumps) or passively vented are a common feature on many landfills and they have been described in detail elsewhere.[41,42]

The construction of composite and multiple liners for leachate control has had, and will continue to have, an increasingly significant affect on landfill gas migration and its control. The use of synthetic liners such as HDPE will considerably decrease the potential for lateral gas migration. Thus, while the hazards associated with gas migration are likely to decrease, the potential for gas utilization may increase.

Landfill Gas Utilization

The exploitation of wastes for energy production has been assessed as being amongst the most economically attractive sources of renewable energy, and

Containment', AEA Technology, Harwell, 1993.
[38] A. Gendebien *et al.*, 'Landfill Gas From Environment to Energy', Commission of the European Communities, Brussels, 1992.
[39] Derbyshire County Council, 'Report on the Non-statutory Public Enquiry into the Gas Explosion at Loscoe, Derbyshire, 24 March 1986', County Offices, Matlock, Derbyshire.
[40] *Ends Rep.*, 1993, **217**, 7.
[41] Department of the Environment, 'Waste Management Paper No. 26A: A Technical Memorandum Providing Guidance on Assessing the Completion of Licensed Landfill Sites', HMSO, London, 1993.
[42] Department of the Environment, 'Waste Management Paper No. 27: Landfill Gas—A Technical Memorandum Providing Guidance on the Monitoring and Control of Landfill Gas', HMSO, London, 1992.

Landfill

recent estimates suggest that the energy equivalent of 26 million tonnes of coal year^{-1} can be derived from waste.[43]

Within the UK, the exploitation of landfill gas as a source of energy has been stimulated under the Non-Fossil Fuel Obligation (NFFO), under which the regional electricity companies are obliged to secure a proportion of their electricity from non-fossil fuel sources. Under this scheme, electricity generated from landfill gas is guaranteed a market at a fixed price, encouraging control of landfill gas. Alternatives such as 'flaring' of the landfill gas, or venting without flaring are wasteful of the gas produced, whilst the latter especially is potentially much more harmful to the environment, through the release of greenhouse gases, as discussed above.

Concerns such as these, together with concerns over the low efficiency of gas collection systems (with resultant high levels of release of fugitive gases—including greenhouse gases and ozone-depleting gases) has fuelled the debate over the use of landfills for the deposit of organic material. Protagonists argue that the generation of energy from waste can be achieved more efficiently and with less pollution through incineration. A recent report[44] concluded that landfill without energy recovery may be responsible for (net) external costs (*e.g.* disamenity affects, damage caused by leachate, pollution caused by transport of wastes to landfill) of £3.4–4 tonne^{-1}, plus disamenity costs. Landfill with energy recovery may be responsible for (net) external costs of £1.3–2 tonne^{-1}, plus disamenity costs, while incineration of waste may lead to (net) external benefits, less disamenity costs (*e.g.* the external benefits may accrue through the replacement of more polluting electricity generation methods with energy derived from the incineration of wastes). On this basis, the incineration of waste, with energy recovery, would appear to be the preferred option. A review of energy to waste issues and examples of schemes currently in progress are available elsewhere.[45]

Landfill Gas Monitoring

At the time of writing, a range of specifications for the monitoring of landfill gas is identified within the EU landfill Directive. These are highly prescriptive, and for many landfills will represent a significant increase in the extent and frequency of monitoring that is undertaken—with an associated increase in costs. In general, the design, frequency, extent, and type of landfill gas monitoring system will be determined by a range of factors, including the amount, composition and rate of gas produced, the location of sensitive targets, the nature of the surrounding geology, and the requirement of statutory guidance and legislative pressures.

Landfill gas measurements, be they from deep monitoring boreholes or surface soil measurements, taken at a single point in time reveal little about the gas

---

[43] Department of the Environment, 'Waste Management Paper No. 1: A Review of Options', HMSO, London, 1992.
[44] Department of the Environment, 'Externalities from Landfill and Incineration', HMSO, London, 1993.
[45] S.J. Burnley and A. Van Santen, 'Proceedings 1992 Harwell Waste Management Symposium: Options for Landfill Containment', AEA Technology, Harwell, 1992.

regime in and around a landfill. It has become clear that a range of environmental factors including rainfall and atmospheric pressure, affect the production and movement of landfill gas and that measurement at a single point in time cannot account for variations caused by the above.[46] For many landfill sites, seasonal trends in gas composition and migration can be detected, and in order to begin to understand the 'gas regime' it would be necessary to monitor such sites for at least one year, after which time the analysis of trends in gas composition and flow will be of more value than the individual measurements. Exceptions to this would be when, at a single point in time, the gas composition and flow rate were particularly hazardous.

When monitoring landfill gas, it is particularly important to understand how the monitoring instruments work and, for example, to recognize the difference between those meters that specifically measure methane and those that measure 'flammable gas'. For any landfill gas monitoring scheme, it is important to recognize why the monitoring is being undertaken and what the objectives are. The answer to these questions will help to effectively design and implement a monitoring scheme that is appropriate to the local conditions.

Detailed guidance on landfill gas monitoring is available elsewhere.[42,47]

6 Site Closure and Aftercare

An increasing awareness of the long-term liabilities associated with landfill have led to a greater need for control of the landfill after 'closure', *i.e.* upon completion of the operational phase. Within the UK, the importance of the post-operational phase has been recognized under Section 61 of the Environmental Protection Act [EPA (1990)], although this particular section has not been implemented so far. If implemented, Section 61 requires that 'it shall be the duty of every waste regulation authority to cause its area to be inspected from time to time to detect whether any land is in such a condition, by reason of the relevant matters affecting the land, that it may cause pollution of the environment or harm to human health'. Where land is found to have potential to cause pollution of the environment or harm to human health, it becomes the duty of the regulation authority to inspect the land in order to keep the condition under review. Also, where pollution is likely, it becomes the duty of the authority to remediate that land, whereupon the cost can be recovered from the owner. However, where the authority has accepted the surrender of a site licence under Section 39 of the EPA (1990), they are not entitled to recover costs of remediation. It is clear that the licensing authority will therefore need good evidence that a site is no longer likely to cause pollution of the environment or harm to human health. If this is accepted, then the authority will issue the owner with a 'certificate of completion'. To help in determining whether or not a completion certificate should be issued, the UK Department of the Environment have issued statutory guidance in the form of Waste Management Paper 26A (Landfill Completion).[41] This paper identifies leachate criteria that should be satisfied before a waste management

[46] Department of the Environment, 'The Technical Aspects of Controlled Waste Management: Understanding Landfill Gas', Report CWM/040/92, 1992.
[47] Institute of Wastes Management, 'Monitoring of Landfill Gas', 1991.

Landfill

Table 6 Completion criteria for leachates

Determinand		Concentration[1]
pH		6.5–8.5
Conductivity		4000
Chloride	as Cl	2000
Sulfate	as SO_4	2500
Calcium	as Ca	1000
Magnesium	as Mg	500
Sodium	as Na	1500
Potassium	as K	120
Aluminium	as Al	2
Nitrate	as NO_3	500
Ammonia	as NH_4	1
Total Organic Carbon	as C	5
Iron	as Fe	10
Manganese	as Mn	2
Copper	as Cu	0.5
Zinc	as Zn	1
Phosphorus	as P_2O_5	1
Fluoride	as F	10
Barium	as Ba	10
Silver	as Ag	1
Arsenic	as As	0.1
Cadmium	as Cd	0.5
Cyanides	as CN	0.05
Chromium	as Cr	0.5
Mercury	as Hg	0.5
Nickel	as Ni	0.01
Lead	as Pb	0.5
Antimony	as Sb	0.5
Selenium	as Se	0.1
Mineral Oils		0.1
Phenols	as C_6H_5OH	0.1
Organochlorine		0.005
Compounds other than pesticides		0.01
Pesticides—individually		0.001
Pesticides—collectively		0.005
PAHs		0.002

[1] All values are in mg l^{-1} except pH (units) and Conductivity (u S^{-1} cm^{-1}).
Source: Waste Management Paper 26A.[41]

license can be surrendered and are based on the Drinking Water Directive (80/778/EEC),[48] assuming at least ten-fold dilution of the leachate within an aquifer (Table 6). In this document, 'completion' is defined as 'that point at which a landfill has stabilized physically, chemically, and biologically to such a degree that the undisturbed contents of the waste are unlikely to cause pollution of the environment or harm to human health (the Completion Condition). At

[48] Commission of the European Communities, 'Directive relating to the quality of water intended for human consumption', 80/778/EEC, OJ L 229, 11.

completion, post-closure pollution controls, leachate management, and gas-removal systems are no longer required'. Six factors are identified that need to be addressed in determining whether or not a site meets the completion condition. They are as follows:

(i) the quality and quantity of leachate present;
(ii) the flow and concentration of gas;
(iii) the potential for polluting leachate or gas to be generated in the future;
(iv) the potential for leachate or gas to reach sensitive targets;
(v) the possibility of physical instability of the waste or retaining structures; and
(vi) the presence of particular problem wastes which could present a hazard in the future.

WMP 26A identifies relevant determinands, sample spacing, and monitoring frequencies. For example, completion criteria for leachate entering groundwater where a minimum dilution factor of ten is expected are identified and are based upon levels in the Drinking Water Directive (80/778/EEC).

Further guidance relating to completion certificates for all sites, including those other than landfill sites, and monitoring guidance during the operational phase of a landfill is given in WMP 4.[49] WMP 4 also emphasizes that site restoration and aftercare is a consideration of the planning authority and as such will be an issue at the planning stage where land usage and the type of restoration will be considered. The final end use of the site will vary according to the local environmental conditions. Technical issues concerned with, amongst other things, the provision of the suitable soil cover, cap protection, soil engineering, and species selection have been discussed elsewhere.[50]

7 Costs, Financial Provision, and Liabilities

The increased costs associated with the monitoring requirements identified in WMPs 4 and 26A, the ever-more stringent engineering requirements, and a range of other costs associated with implementation of new landfill licensing controls are placing greater financial pressure on the management and operation of landfills. Recognizing this, the requirement for adequate financial provision is identified as one of the essential factors in determining a 'fit and proper person' as defined in the EPA (1990) Section 74. Subsection [3(c)] provides that an applicant for a waste management license should be treated as not being a fit and proper person if it appears to the regulation authority that

(i) he has not made financial provision adequate to discharge the obligation arising from the license, and
(ii) he either has not intention of making it, or is in no position to make it.

The effectiveness of this section has yet to be determined; however, at the time of writing, the EU Landfill Directive places further financial requirements on landfill

---

[49] Department of the Environment, 'Waste Management Paper No. 4: Licensing of Waste Management Facilities', HMSO, London, 1994.
[50] Department of the Environment, 'Waste Management Paper No. 26: Landfill Wastes', HMSO, London, 1986.

operators. The Directive requires that all sites, including those in existence before the introduction of the relevant sections of the EPA (1990), should provide evidence of adequate financial provision. Failure to do so would bar the competent authority from issuing a permit to operate. These provisions within the Landfill Directive could cause the closure of many landfill sites across Europe within the next ten years. WMP 4 identifies activities for which adequate financial provision will be required and suitable means of financial provision for each.

Issues relating to insurance for both gradual and sudden and accidental pollution have been discussed elsewhere.[51] The attainment of suitable insurance, or some form of cover against long-term pollution incidents, will be required in the UK to assure 'adequate financial provision'. However, because of the long-term pollution potential of landfill, and the extent of pollution/harm that could occur (*e.g.* the cost of clean-up of an aquifer pollution by landfill leachate has been estimated to run into millions[52]), such cover is currently both difficult and expensive to obtain.

As well as placing greater financial pressures on waste management companies, the increasing controls on the management and operation of landfill sites is causing the cost of waste disposal via this route to increase. For example, within the UK, the National Association for Waste Disposal Contractors (NAWDC) has suggested[53] that the implementation of the new licensing laws will increase the cost of disposal to landfill by £4–5 tonne^{-1}.

8 Landfill—The Future?

The design, management, and control of landfills has changed considerably in recent years in response to a better understanding of the pollution potential of wastes and of more effective means of control.

'Sustainable' and 'fail-safe' concepts are based upon landfills that continue to accept mixed MSW, but in full recognition of the fact that long-term pollution exists. Control may be achieved either by accelerating the rate of waste degradation to produce a stable residue, or by pre-treating the waste prior to landfill. In a recent report, the findings of a feasibility study to assess the potential of developing a bioreactor cell rotation landfill in which both options are utilized have been presented.[54] This scheme differs from previous schemes, such as 'Landfill 2000',[55] in that it incorporates waste separation with homogenization of waste prior to emplacement, in this way 'providing a predictable gas generation rate and a high quality residue as part of a controlled process'. The removal of recoverable materials such as metals, plastics, glass, paper, and textiles ensures that only low calorific value (CV) wastes are landfilled, while high CV wastes are incinerated with energy recovery.

[51] H. Pearce, *Land Contam. Reclam.*, 1993, **1**, 172.
[52] *Ends Rep.*, 1994, **230**, 35.
[53] *Ends Rep.*, 1994, **229**.
[54] Applied Environmental Research Ltd., 'The Sustainable Landfill: A Feasibility Study to Assess the Potential of Developing a Bioreactor Cell Rotation Landfill', ETSU Report B/B3/00242/REP, Harwell, Oxfordshire, 1993.
[55] K. Bratley and Q. Khan, 'Proceedings AMA Environment Conference "Caring for the future"', Newcastle-upon-Tyne, 1989.

In keeping with this theme, Kent County Council (in South-East England) have launched a waste management strategy to reduce the proportion of MSW (650 000 tonnes) and Civic Amenity (CA) waste that goes to landfill. Proposals[56] included an anaerobic digestion plant to handle 40 000 tonnes annum^{-1}, with 200 000 tonnes annum^{-1} going to a waste-to-energy incineration plant.

The option proposed by Kent County Council is similar to that described above but using more traditional anaerobic digestion of the organic waste component rather than the 'bioreactor cell rotational landfill'. Anaerobic digestion (AD) plants such as the DRANCO and VALORGA processes have been reviewed elsewhere.[57] While it is clear that anaerobic digestion after pre-sorting of wastes is receiving more interest, a number of uncertainties relating to the capacity, reliability of the processes, and the marketability of the final product, are still to be resolved. For example, it is possible that if a market cannot be found for the final product of AD, the cost of pre-treatment, together with potential costs associated with segregation and collection at source, may be too high to sustain as a permanent waste management option. Also, if no market can be found for the final product, disposal to landfill would still result in the production of landfill gas and leachate and the benefits of pre-treatment would be negated to a large extent. Thus, while in theory schemes such as those described above seem both feasible and practicable, they have yet to be proven and the full ramifications may not have been realized. The composting of waste (aerobic digestion, not anaerobic digestion) is also receiving increasing interest as a waste treatment option, but at the moment uncertainties similar to those for anaerobic digestion are restricting its use on a commercial scale. The success of these pre-treatment processes may well be dependent upon the willingness of the public to **effectively** sort waste 'at source' and to ensure, for example, that glass is not allowed to contaminate sorted vegetable material.

Even if the products of AD and composting are marketable, landfill will still be required for the final disposal of wastes (ash) from incineration (approximately 15% by weight of input). Such wastes may be landfilled directly, although there is an argument that for the Best Practicable Environmental Option (BPEO)[58] to be achieved, these wastes, which may be high in heavy metal concentration, should be further treated to minimise the pollution risk. This will be especially true if organic wastes are removed from wastes to landfill. In this event, pH control of the environment (achieved through bacterial processes in organic waste degradation) will be significantly reduced, thereby increasing the risk of metal migration to the surrounding environment.

The removal of organic wastes from landfill will also have important ramifications for co-disposal, for without organic material there will be no substrate for growth of the bacteria that effect the degradation of co-disposed wastes. Thus, even if co-disposal is allowed as a waste management option in the future, practical problems may prevent its use.

[56] *Warmer Bull.*, 1993, **39**, 2.
[57] J. Coombs, 'The Prospects for Methane Recovery from the Anaerobic Digestion of Municipal Solid Waste in the UK', ETSU Report B1234S, Harwell Laboratory, Oxfordshire, 1990.
[58] Royal Commission on Environmental Pollution, 'Twelfth report: 'Best Practicable Environmental Option', 1988, Cmd. 310, HMSO.

9 Conclusion

At a time of great change, the future direction of landfill remains uncertain and difficult to predict. Landfill undoubtedly is, and will continue to be, an essential waste management option. However, it seems likely that current changes in regulation and control, and their economic ramifications will, together with political factors and social pressures, lead to changes in waste management as a whole. It is clear that there is an increasing shift in policy towards the top of the waste hierarchy as identified in the EU Fifth Environment Action Programme. For example, in the Netherlands[59] a target of 10% reduction in waste by the year 2000 has been set. There will also be a reduction in the amount of waste disposed to landfill and incineration while at the same time sending a smaller proportion to landfill and a greater proportion to incineration. Policy at both national and regional levels is currently shifting towards the incineration of high CV waste, with energy recovery. Although options for the pre-treatment of lower CV organic wastes are relatively poorly developed on a commercial scale, there is increasing interest in these methods of waste management. This suggests that the landfill of the future will have an extremely low organic material content. Thus, irrespective of political decisions and scientific suitability, processes such as co-disposal are likely to be considerably restricted in the future, unless undertaken through the use of highly controlled and engineered reactors.

Under this scenario, financial provision for effective control of operational and post-operational phases of landfill could be considerably easier to achieve, and the financial elements of landfill generally may be more easily managed.

---

[59] J. P. V. M. Laurijssens, *Wastes Manage.*, 1993, Nov, 12.

Emissions to the Atmosphere

G. H. EDULJEE

1 Introduction

All waste management activities have the potential to release emissions to atmosphere, perhaps along with releases to other environmental media such as groundwater or surface water. These emissions may be *controlled* (*i.e.* managed so as to minimize harm to the environment) or *uncontrolled* (*i.e.* not under the direct management of plant or site operators).[1] In some waste management processes (for example incineration, gasification, pyrolysis) continuous and controlled emissions to atmosphere are an integral and essential consequence of the treatment, since the aim is to convert wastes to essentially gaseous, non-toxic products which can safely be released to the environment. In other waste management processes (physicochemical treatment, solidification, chemical recovery, *etc.*) the aim is not to permit releases to the atmosphere, and to contain the products within the reaction vessel. However, adventitious, uncontrolled releases could still occur if the reactions are not adequately monitored. In the landfilling of biodegradable wastes, approximately 6 m^3 of gas is released per tonne of waste deposited per year. Two gas management concepts are currently under discussion: maintaining dry conditions within the landfill by preventing the ingress of air and water and thus slowing the generation of landfill gas and leachate (the 'dry tomb'), or treating the landfill as a bioreactor and maximizing the generation and release of landfill gas by optimizing air and water requirements.[2]

In addition to these process-related emissions, uncontrolled releases to atmosphere could occur during any stage of the waste management cycle, for example during handling, transportation, inter-plant transfer, *etc.* Figure 1 illustrates the potential for releases from various types of waste management processes: landfills (representing an area source), combustion plant (stack emissions representing a point source), and traffic (representing a line source). Emissions can be continuous (as in the case of gases from incinerator stacks or of gas diffusing through the cover material of a landfill) or discontinuous (as in the

---

[1] M. D. LaGrega, P. L. Buckingham, and J. C. Evans, 'Hazardous Waste Management', McGraw Hill, New York, 1994.
[2] R. W. Maurer, 'Proceedings of the Conference "Landfill Tomorrow—*Bioreactors or Storage*"', Imperial College Centre for Environmental Control and Waste Management, London, 1993.

Figure 1 Emissions to atmosphere from various types of waste management operations, and uptake of contaminants from the environment

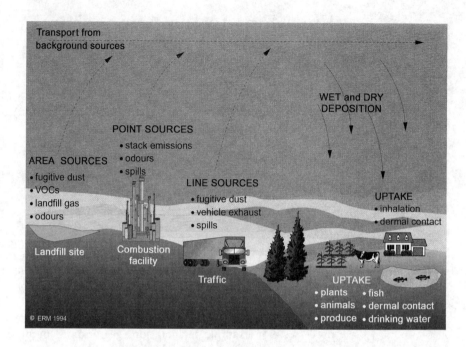

case of releases from physicochemical treatment plants or from fires and accidental spills).

In comparison to soil and ground or surface water, the atmosphere is a far more effective carrier of pollution in that dispersion of emissions into the surrounding environment is multidirectional, relatively fast, and over longer distances.[3] To persons living in the vicinity of waste treatment and disposal facilities, emissions to atmosphere are often the most obvious manifestation of waste management operations, because of visible emissions (steam and smoke from an incinerator stack), or because fugitive releases of dust, odours, litter, *etc.* have resulted in loss of amenity. Certainly, waste management processes that routinely discharge to atmosphere (in general, all thermal processes) have attracted more attention from environmental groups and the public due to anxieties over serious adverse health effects than have other types of waste treatment and disposal options.

This article presents an overview of emissions to atmosphere from waste management operations. The discussion commences with a brief review of the regulatory control of such emissions, as it pertains in the UK. Next, the releases to atmosphere are characterized in terms of their sources, composition, and scale of release. The environmental impacts of such releases are then examined, differentiating between national or global issues, and near-field effects which have the potential to produce more local impacts. Finally, measures which can be taken to mitigate the potential adverse effects of emissions to atmosphere are discussed.

[3] J. Petts and G. H. Eduljee, 'Environmental Impact Assessment for Waste Treatment and Disposal Facilities', John Wiley, Chichester, 1994.

2 Regulatory Framework in the UK

Waste management processes in the UK can be separated into *prescribed processes* and *non-prescribed processes*. Each type of process falls under a different regulatory regime, as discussed below.

Prescribed Processes

The regulatory control of emissions from prescribed processes is effected through the Environmental Protection Act (EPA) 1990, which introduced the concept of Integrated Pollution Control (IPC). Under IPC, pollution of air, water, and land resulting from emissions of 'prescribed' and other substances is regulated within a single framework by the application of two principles, set down in Sections 7(2)(a) and 7(7) of the Act:

- The waste management process which is selected for a particular waste stream must represent the Best Practicable Environmental Option (BPEO), *i.e.* the option which, for a given set of objectives, provides the most benefit or least damage to the environment as a whole, at acceptable cost, in the long term as well as in the short term.[4]
- It must be demonstrated to the satisfaction of the regulatory authorities that process technology, plant, and operating and management regimes have been selected by application of Best Available Techniques Not Entailing Excessive Cost (BATNEEC) to prevent, or, if not practicable, to reduce to a minimum and render harmless, both prescribed and other substances being released to air, land, and water.

The IPC system operates by expressing the performance of designated technologies (representing *Best Available Techniques*) in terms of maximum allowable release levels for prescribed substances against which the actual performance of the prescribed process can be regulated and controlled. Sanctions can then be applied by the regulatory authority if these release levels are breached. BATNEEC, in concert with BPEO, represent the most proximate form of regulatory control since they address (and aim to optimize) the choice of process, plant, equipment, and management regime before emissions are permitted to be released to the surrounding environment, including the atmosphere.

SI 472 (1991)[5] contains a schedule of prescribed processes, divided into Part A and Part B processes. The operations, emissions, and discharges from Part A processes are regulated in their entirety in the UK by Her Majesty's Inspectorate of Pollution (HMIP) through the issuance of an Authorization. The regulation of waste management operations and discharges from Part B prescribed processes is divided between Local Authorities (for emissions to atmosphere), the National

---

[4] Royal Commission on Environmental Pollution, 'Air Pollution Control: An Integrated Approach', Fifth Report, HMSO, London, 1976.
[5] The Environmental Protection (Prescribed Processes and Substances) Regulations 1991, SI No. 472, as amended by The Environmental Protection (Prescribed Processes and Substances) (Amendment) Regulations, 1992, SI No. 614.

Table 1 Prescribed waste treatment processes, controlled by HMIP (Part A) or the Local Authority (Part B)

Part A	Subject
IPR 1/4	Waste and recovered oil burners, >3 MW thermal
IPR 1/5	Combustion of solid fuel manufactured from municipal waste, >3 MW thermal
IPR 1/6	Combustion of fuel manufactured from or comprising tyres, tyre rubber, or similar rubber waste, >3 MW thermal
IPR 1/7	Combustion of solid fuel manufactured from or comprising poultry litter, >3 MW thermal
IPR 1/8	Combustion of solid fuel which is manufactured from or comprises wood waste or straw, >3 MW thermal
IPR 5/1	Merchant and in-house chemical waste incineration
IPR 5/2	Clinical waste incineration, >1 t h^{-1}
IPR 5/3	Municipal waste incineration, >1 t h^{-1}
IPR 5/4	Animal carcasses incineration, >1 t h^{-1}
IPR 5/5	Burning out of metal containers
IPR 5/6	Making solid fuel from waste
IPR 5/7	Cleaning and regeneration of carbon
IPR 5/8	Recovery of organic solvents by distillation
IPR 5/9	Regeneration of ion exchange resins
IPR 5/10	Recovery of oil by distillation
IPR 5/11	Sewage sludge incineration, >1 t h^{-1}
Part B	**Subject**
PG 1/1	Waste oil burners, <0.4 MW thermal
PG 1/2	Waste or recovered oil burners, <3 MW thermal
PG 1/6	Tyre and rubber combustion, 0.4–3 MW thermal
PG 1/7	Straw combustion processes, 0.4–3 MW thermal
PG 1/8	Wood combustion, 0.4–3 MW thermal
PG 1/9	Poultry litter combustion, 0.4–3 MW thermal
PG 5/1	Clinical waste incineration, <1 t h^{-1}
PG 5/2	Crematoria
PG 5/3	Carcass incineration, <1 t h^{-1}
PG 5/4	Waste incineration, <1 t h^{-1}
PG 5/5	Sewage sludge incineration, <1 t h^{-1}

Rivers Authority (for discharges to surface waters), the sewerage undertakers (for discharges to sewer), and the Waste Regulation Authority (control of storage, handling, and operations on site). Table 1 lists the prescribed waste management processes together with a reference to the relevant Process Guidance Notes. In addition to providing guidance on technologies and practices that currently represent Best Available Techniques, these documents also stipulate the release levels for prescribed substances, that prescribed processes cannot exceed when discharging to atmosphere.

A fuller discussion of the IPC system, BPEO, and BATNEEC has been presented in Chapter 1.

Non-prescribed Processes

The regulation of emissions under the IPC system is based on the premise that the appropriate technology and operational practice adopted for a particular waste management operation can be translated into a set of emission limits, which in turn provides the focus for regulation. The waste management processes which are most amenable to this approach are those which release controlled, continuous emissions and which utilize plant and equipment similar to that encountered in industrial processes. In general, all thermal processes fall into this category. However, incineration accounts for only a small percentage of wastes which are dispatched for disposal in the UK: 7% of municipal solid waste (MSW), and about 5% of industrial wastes. The predominant form of disposal for both municipal and industrial wastes is landfill (about 90% of municipal and 70% of industrial wastes) followed by other types of treatment and disposal methods such as physicochemical treatment, composting, anaerobic digestion, *etc*. Although all of these latter processes have the potential to release emissions to atmosphere, none are classed as prescribed processes, and hence they fall outside the remit of IPC.

Other regulatory control mechanisms are available to ensure that emissions from non-prescribed processes do not cause adverse environmental effects. Chief among these is the licensing of waste management sites by Waste Regulation Authorities, through which performance objectives can be specified. These generally take the form of statements requiring the site operator to manage the process such that no detrimental effects are manifested beyond the site boundary, coupled with suitable monitoring arrangements. Guidance on landfilling of wastes is additionally provided in the series of Waste Management Papers published by the Department of the Environment, in particular Waste Management Paper No. 26 on landfill design and practice,[6] and Waste Management Paper No. 27 on the control of landfill gas emissions.[7]

A further safeguard against inappropriate choice of technology and operating methods is afforded through the UK planning regime, for most new waste management plants or for existing sites undergoing significant change. Such operations will generally require an Environmental Assessment to be performed and subjected to scrutiny by regulators and the public.[3] Impact on the environment is a material planning consideration, requiring developers to demonstrate that the intended operation (including any emissions to atmosphere) will not have a detrimental effect. The choice of appropriate technology and operating practices underlies any assumptions regarding the benign nature of the resulting emissions to atmosphere.

EU Legislation

The European Union (EU) has introduced a number of Directives, elements of which have been incorporated into UK legislation relating to emissions from waste management processes. Directive 84/360/EEC on the Combating of Air Pollution from Industrial Plants introduced the concept of Best Available Technology 'that does not entail excessive costs' to prevent or reduce air

---

[6] Department of the Environment, 'Waste Management Paper No. 26: Landfilling Wastes', HMSO, London, 1986. (Currently under revision).
[7] Department of the Environment, 'Waste Management Paper No. 27: Landfill Gas', HMSO, London, 1991.

pollution: the UK's incorporation of BATNEEC in the EPA 1990 (with a wider interpretation inherent in the substitution of *Techniques* for *Technology*) is a consequence of the implementation of this Directive.

The EU also recognized in the Fifth Environmental Action Programme the need to ensure high and consistent waste management standards across the Union, and to this end are in the process of issuing Directives addressing various waste treatment operations. Directives 89/369/EEC and 89/429/EEC introduced standards of design and operation for municipal waste incineration. Draft Directives on hazardous waste incineration[8] and on landfilling[9] are under discussion; the former incorporates more stringent emission limits than those which currently apply in the UK through Process Guidance Note IPR 5/1, while the draft Directive on landfilling incorporated performance objectives in respect of the control of landfill gas emissions. Once issued, the substance of the Directives, including the stipulated emission limits, is then incorporated into the legislation of each Member State. The standards introduced in the Directives are under constant review. For example, it has been proposed by the EU that the incineration of all materials classified as wastes should be subject to common emission limits.[10]

3 Characterization of Emissions

Releases to atmosphere can occur at every stage of waste transport, handling and treatment. For example, during transport the potential exists for the following emissions to occur:

- *Dust*; from tracking of contamination on the wheels of the vehicle.
- *Dust*; from the payload, from dry and finely dispersed material that has not been sufficiently damped or sheeted.
- *Odours*; from open loads, leaking valves, biological or organic sludges.
- *Adventitious releases* of gases or fumes as a result of accidents and fires.

Similarly, during landfilling of wastes, releases to atmosphere can be characterized as follows:

- *Dust*; tracking of vehicles on the site, working of the landfill, and of the cover material.
- *Litter*; from loose, unbagged waste and uncovered working areas.
- *Odours*; landfill gas, decomposing waste, chemicals.
- *Pathogens and microbial emissions*; release of bacteria and pathogens through disturbances of the waste during landfilling and compaction.
- *Combustion gases* from the flaring or utilization of landfill gas.
- *Adventitious fumes* from fires, gas explosions, and uncontrolled chemical reactions.

[8] Commission of the European Communities, 'Proposal for a Council Directive on Hazardous Waste Incineration', COM(92)9 Final-SYN 406, *Off. J. Eur. Communities*, C130, 21 May 1992.
[9] Commission of the European Communities, 'Proposal for a Council Directive on the Landfill of Waste', (91/C190/01), *Off. J. Eur. Communities*, C190/1-18, 22 July 1991. Amended proposals COM(93)275, 1993.
[10] Commission of the European Union, 'Draft Directive on Incineration of Waste', 21 April 1994.

Emissions to the Atmosphere

Table 2 Relative contribution of stack and fugitive emissions to atmosphere from a hypothetical waste incineration facility[11]

Chemical	Emissions/kg y^{-1}		Contribution to glc*/%	
	Stack	Fugitive	Stack	Fugitive
Chloroform	11	42	15	85
Ethylene dichloride	35	47	32	68
Hexachlorobutadiene	17	0.2	98	2
Tetrachloroethane	4	0.6	84	16
Arsenic	4	0	100	0
Chromium	17	0	100	0
Lead	22	0	100	0
Phenol	7×10^{-3}	1×10^{-5}	98	2
Toluene	39	18	58	42
Pyridine	3	1	64	36
Phthalic anhydride	2×10^{-3}	1×10^{-5}	99	1
Methyl styrene	37	3	90	10

*glc = ambient air ground level concentration at point of maximum impact.

Emissions sources from waste management processes such as incineration and physicochemical treatment are similar to those encountered in the chemical and process industries; namely, process vents and stacks, scrubbers, *etc*. In addition to deliberate and controlled discharges (such as via a tall stack), fugitive emissions can occur from pumps, valves, seals, waste handling and transfer operations, opening and sampling of drums, displacement of headspace in storage tanks, *etc*. The cumulative magnitude of these fugitive releases has often been overlooked when compiling an inventory of releases from a site. In one study of an incineration facility[11] an attempt was made to characterize fugitive release from waste handling and storage operations associated with rotary kiln and liquid injection combustion systems fed with pesticide wastes, oily sludge, and phenol/acetone distillation wastes. Table 2 summarizes the annual emissions of representative chemicals in the wastes streams from the stack and from fugitive sources during the operation of a hypothetical medium-sized liquid injection combustor. It can be seen that fugitive releases can reach significant proportions relative to emissions from the stack and, for the more volatile chemicals, can be the main contributor to ambient air ground level concentrations.

In another study, annual emissions of five metals from the stack of an MSW incinerator were compared against annual fugitive emissions of these metals in dust released during the landfilling of the ash residue.[12] For the landfill scenario, it was assumed that dust emissions derived from three sources: placement of the ash (including transport over the landfill), spreading and compaction of the ash, and wind erosion. The latter source accounted for approximately 75% of the metals emitted from the landfill. However, the annual emissions of metals from the stack exceeded the annual fugitive emissions from the landfill by a factor of 50–500.

[11] C.C. Travis, E.L. Etnier, G.A. Holton, F.R. O'Donnell, D.M. Hetrick, E. Dixon, and E.S. Harrington, 'Inhalation Pathway Risk Assessment of Hazardous Waste Incineration Facilities', Report ORNL/TM-9096, Oak Ridge National Laboratory, Oak Ridge, Tennessee, 1984.

[12] D.A. Kellermayer and S.L. Stewart, *Environ. Impact Assess. Rev.*, 1989, **9**, 223.

Odour is perhaps the most common type of fugitive emission released from waste treatment or disposal sites: this is reflected in the predominance of complaints relating to odour nuisance over other types of disamenity which could potentially affect surrounding populations. In most waste management processes and operations associated with handling of waste, odours are released as a result of ineffective or nonexistent fume extraction and odour control measures that ideally should be engineered into the design of the plant or incorporated into the systems of work. In processes such as composting, the act of aeration and/or turning of the waste to encourage microbial activity leads to the generation and release of odours, the intensity of which depends on the state of decomposition of the waste and the method of composting. In one study[13] odour emissions for different composting techniques were characterized in terms of Odour Units (OU). The discharges ranged from 1000 OU m^{-3} before turning of the compost, to 5000 OU m^{-3} during turning of a 4 weeks old pile. Blowing of air through the bottom of the pile resulted in lower odour emissions (200 OU m^{-3}) than when air was drawn from the top of the pile (20 000 OU m^{-3})—in the former case, the top layer of waste partly deodorized the air stream prior to its dispersal in the surrounding air. Appropriate working methods and engineered control techniques were suggested to ensure that odours were not detected off-site.

The handling, treatment (for example, composting), and landfilling of Municipal Solid Waste (MSW) can also result in microbial emissions to atmosphere. Pathogenic bacteria form about 6% of the total bacteria on waste reaching a landfill or compost plant.[14] Sampling of air prior to composting of MSW indicated concentrations of 10 000–20 000 Gram-negative bacteria per cubic metre of air, with higher concentrations during subsequent dismantling of the compost pile. These emissions have greater implications for worker safety than for adverse health effects off-site.

The principal and trace components of emissions to atmosphere from a range of waste management activities are shown in Table 3. Other than continuous emissions from thermal processes and from landfills, it is difficult to characterize releases to atmosphere from other processes, since they are either of variable quality (for example, from the scrubber or ventilation system of a physicochemical treatment plant) or are fugitive and of intermittent duration, dependent on the nature of the material processed or handled, and on the work practices observed at any particular time.

A detailed characterization of emissions of organic micropollutants from waste incinerators has been presented in a previous volume in this series.[15] Emission factors relating to MSW incineration (μg emitted per tonne of MSW combusted) for polychlorinated dibenzodioxins and dibenzofurans (PCDDs and PCDFs)

---

[13] W. Bidlingmaier, 'Odour—Emissions from Composting Plants', University of Stuttgart, 1990; personal communication.

[14] J. Lacey, P. A. M. Williamson, and B. Crook, 'Microbial Emissions from Composts made for Mushroom Production and from Domestic Waste', AFRC Institute for Arable Crops Research, Harpenden, UK, 1990.

[15] G. H. Eduljee, in 'Issues in Environmental Science and Technology—No. 2 Waste Incineration and the Environment', ed. R. E. Hester and R. M. Harrison, The Royal Society of Chemistry, Cambridge, 1994.

Table 3 Principal and secondary emissions to atmosphere from various waste management operations

Activity	Type	Major components	Trace components	Other emissions
Thermal processes	Stack emissions	Carbon dioxide Water	Carbon monoxide Acid gases Metals Organics	Odours Adventitious releases from spills, fires, etc.
Landfilling	Landfill gas	Methane Carbon dioxide	Volatile and semi-volatile organics	Odours Dust Litter Pathogens
Physicochemical treatment	Scrubber emissions Building ventilation	Air, water, vapour Air	Acid gases Organics Organics	Odours Adventitious releases from spills, fires, etc.
Biological treatment	Methane (from some processes)	Methane Carbon dioxide	Volatile and semi-volatile organics	Odours

and polychlorinated biphenyls (PCBs) have been published.[16] Emission factors for metals released during MSW incineration are available for a range of combustion technologies.[17] Landfill gas entrains over 100 organic compounds in trace quantities, many of which are potentially odorous when present at sufficiently high concentrations.[18,19] According to one study,[20] about 10% of the trace compounds are likely to present odour problems, the most common types being organosulfurs, esters, organic acids, hydrocarbons such as limonene, and alcohols.

4 National and Global Impacts

The impact of waste management activities on a national scale can be assessed by comparing emissions from such processes with emissions from other activities. Emissions of chemicals such as sulfur dioxide, nitrogen oxides, carbon dioxide, and methane are also of relevance to global issues such as acid rain, ozone depletion, and the greenhouse effect. In this context, waste disposal activities that discharge gases to atmosphere on a continuous basis are of importance, *i.e.* combustion and landfilling.

It is generally acknowledged that, save perhaps for emissions of methane from landfills and of certain trace chemicals associated with waste incineration, waste management activities do not make a significant contribution to national inventories of emissions to atmosphere. For example in the UK, emissions of nitrogen oxides from two sources (motor vehicles and power stations) account for 90% of the national budget for these compounds in urban areas in episode conditions. Power stations alone account for 72% of sulfur dioxide emissions. Combustion of coal, motor spirit, and gas accounted for 75% of the carbon dioxide emitted into the UK atmosphere.[21] In Germany, carbon dioxide emitted from municipal solid waste (MSW) incinerators accounted for less than 1% of total emissions from all sources. Emissions of volatile organic compounds (VOCs) from waste incinerators contributed 0.1% to the total atmospheric burden in Germany, the main sources being road traffic (49%) and the use of solvents in industry (38%).[22] An assessment of the contribution of MSW incineration to the US environment[23] indicated that total acid gas and nitrogen oxide emissions amounted to less that 1% of these emissions from utility coal boilers. The carbon monoxide (CO) released from MSW incinerators amounted

---

[16] US Environmental Protection Agency, 'Estimating Exposure to Dioxin-like Compounds', Review Draft EPA/600/6-88/005B, Office of Research and Development, Washington DC, 1992.

[17] F. T. DePaul and J. W. Crowder, 'Control of Emissions from Municipal Solid Waste Incinerators', Noyes Data Corporation, New Jersey, 1989.

[18] J. Brosseau and M. Heitz, *Atmos. Environ.*, 1994, **28**, 285.

[19] P. J. Young and A. Parker, *Waste Manage. Res.*, 1983, **1**, 213.

[20] P. J. Young and A. Parker, *Chem. Ind.*, 1984, **9**, 329.

[21] Department of the Environment, 'First Report of the Quality of Urban Air Review, Review Group', London, 1993.

[22] Ministry for Environmental Affairs, 'Questions on Waste Incineration', Document UM-8-89, Ministry for Environmental Affairs, Stuttgart, 1989.

[23] R. S. Egdall, A. J. Licata, and L. A. Terracciano, in 'Proceedings of the GRCDA/SWANA Sixth Annual Waste-to-Energy Symposium', Arlington, Virginia, 1991.

Table 4 Average global emissions of trace elements in 1983 from a variety of industrial and waste-related activities (tonne y^{-1})[27]

	As	Cd	Cr	Cu	Hg	Ni	Pb
Coal combustion	1980	530	11275	5185	2080	13760	8160
Oil combustion	57	144	1410	1960	—	27100	2420
Pyrometallurgy	12255	5430	—	23260	120	7990	46535
Iron and steel	1420	160	15620	1490	—	3570	7630
MSW incineration	270	730	540	1470	1120	260	2100
Sewage sludge incineration	40	20	300	105	40	105	270
Wood combustion	180	120	—	900	180	1200	2100
Cement production	535	270	1335	—	—	490	7130
Miscellaneous	2025	—	—	—	—	—	4500
Percentage contribution from incineration	*1.7*	*10*	*2.8*	*4.5*	*33*	*0.7*	*3*

*Non-availability of data is indicated by a (—) sign. The percentage contribution from incineration is therefore an upper limit.

to less than 0.05% of CO emissions from motor vehicles. A comparison was made of emissions of 15 carcinogenic organic compounds and 17 non-carcinogenic organic compounds emitted from hazardous waste incinerators in the US against releases reported by industry in the 1990 Toxic Release Inventory.[24] The total mass emissions of all 32 organics from hazardous waste incinerators (100 tonnes) was less than 0.03% of the corresponding releases from industry (388430 tonnes).

Emissions of specific trace metals and organics from waste incineration have been assessed against total releases to atmosphere. In Sweden, MSW incineration was identified as contributing over 50% of total emissions of cadmium and mercury in the mid-1980s, but with progressively tightening incinerator emission standards coupled with a campaign to remove these metals from MSW, this contribution is expected to fall to about 2% by the mid-1990s.[25] In the UK, cadmium and mercury emissions from waste incinerators are believed to represent 15% and 60%, respectively, of the total emissions of these metals to atmosphere.[26] This study is currently being updated by the Department of the Environment.

A global inventory of trace metals emitted to atmosphere in 1983 has been attempted.[27] Table 4 summarizes the average contribution from each source type, for a selection of metals. Incineration (which includes both MSW and sewage sludge incineration) generally accounts for a small percentage of metal emissions to atmosphere save for mercury.

A number of countries have compiled inventories of PCDD and PCDF emissions to atmosphere.[28-31] Waste (in particular, MSW) incineration had,

[24] C. R. Dempsey, *J. Air Waste. Manage. Assoc.*, 1993, **43**, 1374.
[25] C. Porter, *Warmer Bull.*, 1990, **25**.
[26] P. K. Leech, 'UK Atmospheric Emissions of Metals and Halides', Report LR 923, Warren Spring Laboratory, Stevenage, 1993.
[27] J. O. Nriagu and J. M. Pacyna, *Nature (London)*, 1988, **333**, 134.
[28] J. Schaum, D. Cleverly, M. Lorber, L. Phillips, and G. Schweer, *Organohalogen Comp.*, 1993, **14**, 319.
[29] J. de Koning, A. A. Sein, L. M. Troost, and H. J. Bremmer, *Organohalogen Comp.*, 1993, **14**, 315.
[30] H. Fiedler and O. Hutzinger, *Chemosphere*, 1992, **25**, 1487.
[31] S. J. Harrad and K. C. Jones, *Sci. Total Environ.*, 1992, **126**, 89.

Table 5 National inventories of PCDD and PCDF emissions to atmosphere ($g\ y^{-1}$)[28-31]

	US	Netherlands	W. Germany	UK
Incinerators				
MSW	60–200	382	5–432	11
Clinical	500–5100	2.1	5.4	1.7
Chemical	2.6–8.4	16	0.5–72	+
Sewage sludge	1–26	0.3	0.1–1.13	
Combustion				
Coal		3.7	2.9ˣ	12.8
Oil		1.0	1.2ˣ	
Wood	70–1600	12		
Forest fires	300–3000			
Traffic	<8–870	7	0.1–0.4	0.7
Sintering processes (copper)	230–310	26	38–380	

*$g\ y^{-1}$ of PCDDs and PCDFs expressed as International Toxic Equivalents, I-TEQ.
+'a few grammes of TCDD'.
ˣDomestic heating only.

until the mid-1980s, been the focus of attention as the principal emission source, but more recent studies have identified operations such as secondary metal smelting as significant contributors to national budgets of PCDDs and PCDFs. Whereas in the past MSW incineration may well have contributed disproportionately to the national atmospheric burden of PCDDs and PCDFs, the imposition of increasingly stringent emission limits has resulted in a 100-fold reduction in emissions from this source, and similar emission limits or guidelines applied to other waste incineration processes should further reduce the significance of waste-related thermal processes relative to other sources of emissions. Table 5 summarizes information on annual releases in grams of PCDDs and PCDFs to atmosphere, expressed in terms of International Toxic Equivalents, I-TEQs. Inter-country comparisons are not valid, since the national inventories were conducted at different times and the more recent inventories tend to identify new sources not previously suspected of being significant emitters of PCDDs and PCDFs. In the UK, chemical, MSW, and clinical waste incinerators are believed to contribute approximately 40% of the national atmospheric burden of PCDDs and PCDFs[31] but a broader examination of industry could well identify new sources.

The relative merits of incineration *versus* landfilling have been examined in terms of emissions of greenhouse gases. It is estimated that between 17% to 58% of the UK's annual emissions of 5.3 million tonnes of methane derive from landfills, followed by agricultural sources (30%) and coal mining (14%).[32] One tonne of biodegradable waste generates approximately 6 m³ of landfill gas annum⁻¹, the main constituents being methane (55%) and carbon dioxide (45%). Methane is estimated to be 25–30 times more potent than carbon dioxide

[32] Anon, *Environ. Data Services*, 1993, **217**, 7.

Emissions to the Atmosphere

Table 6 Comparison of emissions from conventional power generation sources and from a replacement 45 MW waste-to-energy incinerator (tonne y^{-1})[23]

	42 MW utility	Office building	from landfill	Total, conventional sources	Replacement 45 MW incinerator
Particulates	93	20	0	113	56
CO_2 equivalent	368 000	26 000	550 000	944 000	575 000
CO	30	104	5	139	143
SO_2	1256	73	4	1333	195
NO_x	867	55	41	963	842
HCl	59	2	0	61	93
Hydrocarbons	4	9	52	65	5

as a greenhouse gas. The greenhouse effect of landfilling one tonne of MSW has been calculated as being equivalent to about 4.8 tonnes of carbon dioxide, as opposed to 0.8 tonnes of carbon dioxide released through incineration, a difference in potency of a factor of six.[1]

Comparisons have also been made between the release of carbon dioxide and other chemicals from waste-to-energy plants (*i.e.* power plants using waste materials as fuel) and from conventional power stations. For example, it has been estimated[33] that the offset in carbon dioxide emissions arising from the utilization of waste-to-energy as a replacement for coal is approximately 430 kg tonne^{-1} of MSW, assuming that one tonne of MSW generates 480 kWh and the emissions of carbon dioxide from coal combustion is 1 kg (kWh)$^{-1}$. This is equivalent to a saving of approximately 50% of the carbon dioxide produced by coal-firing for similar amounts of electricity generation. One estimate for the UK equates the energy value of the 30 million tonnes of MSW generated annually to 10 million tonnes of coal equivalent.[34] A similar comparison, but on a more localized scale, has been provided for a proposed waste-to-energy plant in Bridgeport, Connecticut, to replace 10 MW and 35 MW of energy generated from oil and coal, respectively.[23] The waste-to-energy option resulted in a net decrease of 85% in sulfur dioxide emissions, 44% in particulate matter, and 10% in nitrogen oxides. Due to the magnitude of these decreases, they were considered to more than offset increases in the quantities of carbon monoxide (by a factor of 3.5) and hydrogen chloride (by 48%) emissions.

Another interesting comparison has been made between options for development of the Bridgeport site and associated power sources.[23] The options were development of the site as a 12-storey office building, or as a 1500 tonne day^{-1}, 45 MW waste-to-energy plant to replace existing oil and coal fired power stations. In the former case, 42 MW of power was supplied by an oil and coal fired utility, in addition, the office building had an on-site boiler for providing space heating and other services. Further, 1500 tonnes of MSW would need to be landfilled in the vicinity, resulting in the release of landfill gas, offset by the recovery of a proportion of the gas to supply 3 MW of energy. Table 6 summarizes the net

[33] H. F. Taylor, in 'Proceedings of the USEPA/AWMA Second Annual International Conference on MSW Combustion', Tampa, Florida, 1991.
[34] A. Porteous and R. S. Barrett, 'Proceedings of 'Incineration an Environmentally Acceptable Means of Waste Disposal?', Institution of Mechanical Engineers, London, December 1993.

emission impact from the above options. The emission budget for incineration does not take into account employee travel (included in the office building scenario) and transport of waste to and from the facility. An overall reduction of about 60% in emissions of carbon dioxide and about 50% for the total trace pollutants was computed when the incineration option was adopted.

Such comparisons are increasingly being undertaken on a regional or sub-regional scale in order to place waste management options within the wider context of other emission sources or other development options. Inevitably, this will necessitate a study of the interaction between waste management options (for example, between recycling, landfilling, and incineration) and between waste and non-waste related activities, and will form an important aid to regional and national planning.

5 Localized Impacts

Opposition to waste management operations is often characterized by phrases such as the 'Not In My Back Yard' (NIMBY) syndrome, with waste disposal facilities being described along with power stations and nuclear facilities as 'Locally Unacceptable Land Uses' (LULUs). These phrases typify the primary concern voiced by the public against such activities; their local impact. With landfills, the concerns relating to emissions to atmosphere are primarily those of loss of amenity. With incinerators and other combustion-related activities, the main concerns are those of adverse health effects. These two types of impacts are discussed in the following section.

Loss of Amenity

Emissions to atmosphere can cause disruption to the use and enjoyment of the local environment through the formation of mists and fogs (not generally an issue in waste management processes) and through the release of dust and odours. The latter is by far the most common cause of complaint against a waste facility. Examples of wastes that possess the potential for odour release include domestic refuse, biological sludges from food processing industries, abattoirs, tanneries, *etc.*, and organic chemicals such as mercaptans. Some example of stages during handling, treatment, and disposal which may give rise to odours are as follows:

- off-loading from a tanker;
- discharge from the off-loading pump into a storage or treatment vessel;
- opening of drums and containers in an uncontrolled work environment;
- degradation of waste in a landfill, exacerbated by lack of adequate daily cover;
- forced aeration or raking of compost; and
- emissions of landfill gas along with entrained odorous chemicals.

Stack emissions from combustion processes are rarely a source of odours; they are predominantly from ground level sources such as those described above. Wastes which are identifiable as odorous are generally handled by different methods from other wastes and, in extreme cases following persistent and

Emissions to the Atmosphere

Table 7 Breakdown of nuisance complaints relating to the operation of a physicochemical treatment plant[35]

Year	Total	Odour/taste	Noise	Dust/fume
1	178	178	No records	No records
2	70	70	No records	No records
3	100	85	12*	3*
4	86	35	34	17
5	146	113	10	23

*Partial records for 4 months of the year.

long-standing complaints against the facility, they are excluded from the site through a condition imposed in the site licence.

The composition of odorous emissions can be extremely complex. For example, wastes from animal rendering plants are likely to contain 40–50 separate compounds, each with a distinctive odour and with odour detection thresholds in the range 0.1–10 p.p.m. The nuisance potential of these odours is dependent on the efficacy of the operating practices and the odour control measures designed into the facility. The rate at which odours are emitted from the facility must be low enough for the diluting effects of atmospheric dispersion to bring concentrations at the receptor below the detection threshold. The impact of odours is determined by very short-term peak concentrations, of the order of seconds. Concentrations determined over longer averaging times (say, one-hour averages) may be below the detection threshold of a particular odour, but such an average will be composed of many short peaks several times greater than the overall average. Most complaints are likely to be received when atmospheric dispersion is poor and wind speeds are low. In terms of stability categories, the frequency of occurrence of those categories which would result in the highest concentrations in the UK is approximately 20–25%, composed of 6–7% unstable conditions (hot, sunny weather), and 15–18% stable conditions (cold, cloudless conditions at night or in wintertime).[35] Complaints of odour nuisance are generally not received until the odour concentration perceived by the receptor is approximately five times that of the detection threshold.

Complaints relating to loss of amenity generally comprise a mixture of nuisance from noise, odours, dust, *etc.* They are difficult to interpret because of their intermittent nature, lack of rigour in regulatory investigation and follow-up, and poor record keeping. Further, reactions to potential nuisance are very varied: although an entire neighbourhood may experience discomfort, only one or two residents may be moved to register a complaint, and a small number of persistent complainants can contribute disproportionately to the total number of complaints received. An analysis of complaints against a chemical waste treatment plant illustrates the issues.[35] A breakdown of the local authority records over five years of operation is shown in Table 7.

For Years 1 to 3, local authority records were incomplete, with potentially significant under-reporting of noise and dust/fume complaints in Year 3, and no available records for these emissions in Years 1 and 2. The number and

---

[35] Environmental Resources Management, unpublished data, 1989.

Table 8 Number of days and percentage of working days linked with complaints relating to the operation of a physicochemical treatment plant[35]

Year	Odours Days	%	Noise Days	%	Dust/Fume Days	%
1	101	31	No records	No records	No records	No records
2	53	16	No records	No records	No records	No records
3	64	19	9*	3	2*	1
4	24	7	28	8	16	5
5	60	18	9	3	19	5

*Partial records, for 4 months in the year.

percentage of working days linked with complaints is shown in Table 8, and taken at face value suggests that the plant was a significant source of nuisance. For the purposes of the analysis, 97 out of a total of 580 complaints had to be rejected because some complaints were not accompanied by complaint forms, some complaint forms were illegible, or some complainant addresses could not be located on the local area map. Of the remaining 483 complaints, 55% originated from three complainants, of which two complainants accounted for 67% and 39% of all complaints in Year 4 and Year 5 respectively. Between Year 3 and Year 5, a single resident accounted for 182 complaints while the remaining 150 complaints were distributed between 60 other individuals.

Complaints concerning odours predominated over those for noise and dust/fume. On average, 83% of odour complaints appeared justified in terms of the direction of the prevailing wind. Of the complaints relating to noise, 93% originated from three streets adjoining the works, of which 72% originated from one complainant. On several occasions clusters of complaints (for example, among even-numbered residences on a street) suggested an orchestrated approach to the local authority. One persistent complainant against odour nuisance did not register any complaint against noise nuisance, probably because the pattern of shiftwork at his place of work meant that he was absent during nighttime (when the majority of complaints were lodged) and because his workplace was itself a source of noise. Complaints relating to dust and fume were generally of a visual nature, and often referred to specific emissions from the site, amounting to 5% of operating time or 0.44% of total operating hours, assuming the maximum duration of each event was one hour and the plant operated twelve hours a day. Emission sources on site were at or near ground level, and therefore dust and fumes were likely to affect properties in the immediate vicinity of the plant. All but 4 of the 41 complaints originated from the nearest street, and in particular from the closest receptor, who had the clearest view of the plant and key items of equipment.

The risk of off-flavours in food products exposed to emissions from waste treatment plants has become an increasing concern to food manufacturers, as serious loss of production and customer goodwill could occur in the event of a tainting incident. While there have been no recorded incidents of tainting which could be attributable to emissions from a waste management operation, the level of concern, supported by theoretical calculations pointing to a high probability of tainting following a small release, has been sufficient for this risk to be cited in at least one planning appeal decision against the siting of a chemical waste treatment plant in the vicinity of food manufacturers.[36]

Health Effects

Public anxiety over adverse health effects posed by waste management activities has all but stalled the construction of new and much needed waste treatment and disposal facilities, in particular combustion plant. In Australia, efforts to construct a national hazardous waste incinerator have failed despite a decade of intense activity by Federal and State authorities to convince a sceptical public of the safety of such a plant. In Spain the implementation of the country's national waste plan has been delayed owing to public opposition over the siting of the necessary landfills, physicochemical treatment plants, and incinerators. In Rhode Island, USA, incineration of MSW has been disallowed, and the shortage of indigenous landfill space has necessitated its export to neighbouring States for disposal. The timescale for successful siting of new hazardous waste and major MSW facilities, especially incinerators, is in the order of five years or greater, the majority of proposals being subjected to public scrutiny through the medium of a public inquiry or equivalent.

Emissions to atmosphere constitute the first stage of a number of potential exposure possibilities. The most direct route of uptake into the body is through inhalation of the chemical, or via deposition of contaminated dust onto skin. Indirect exposure pathways result from deposition of the chemical onto soil or water, followed by transfer through the terrestrial foodchain into plants, animals, and finally into humans. Other indirect exposures can occur through the use of contaminated resources, for example:

- drinking or swimming in contaminated water, where uptake into the body is via ingestion and dermal contact;
- showering in contaminated water, where uptake is via dermal contact and inhalation of volatiles stripped from the water due to the action of heat;
- consumption of fish and bottom feeders, where uptake is via bioaccumulation through the aquatic foodchain; or
- direct contact with contaminated soil through activities such as gardening, and pica. Uptake is via dermal absorption and ingestion.

There are a number of potential exposure pathways that contribute to human uptake by ingestion. The relative importance of each exposure pathway depends on the activity patterns of exposed populations, and will therefore differ from site to site. In most scenarios linking exposure with emissions to atmosphere, uptake via ingestion (generally, through the terrestrial foodchain) is the dominant route, with direct routes of inhalation and dermal contact typically accounting for about 10% and 1% to total uptake, respectively. An illustration is provided in Table 9, which summarizes the percent contribution to ingestion from four exposure pathways following deposition onto soil and plants of emissions from a proposed three-stream power plant fired by coal and MSW.[37]

[36] Department of the Environment, 'Decision of the Secretary of State in relation to the local inquiry held to hear the appeal of Leigh Environmental Limited against non-determination of the application for the provision of a regional waste treatment facility', Reference No. APP/F4410/A/89/126733, Department of the Environment, 1991.

[37] Maryland Department of Natural Resources, 'Risk Assessment Study of the Dickerson Site', PPSE-SH-4, 1990.

Table 9 Percentage contribution of four exposure pathways to uptake via ingestion following deposition of emissions from a power plant complex[37]

Pathway	As	Cd	Cr	Pb	Hg	Ni	Dioxins	Formaldehyde	PAH
Soil–human	34	2	32	27	4	1	0.6	0.3	6
Plant–human	59	85	65	48	24	82	12	99	58
Soil–animal–human	5	2	2	19	40	3	59	0	14
Plant–animal–human	2	11	1	6	32	14	29.4	0.7	22

(i) Soil–human: Direct ingestion of contaminated soil.
(ii) Plant–human: Consumption of contaminated plants.
(iii) Soil–animal–human: Consumption of animals that have ingested contaminated soil during grazing.
(iv) Plant–animal–human: Consumption of animals that have eaten contaminated plants.

It was assumed that the facility operated for 70 years (an average human lifespan) and that, following deposition, the metals were not subjected to loss mechanisms such as leaching or volatilization. The study assumed a soil ingestion rate of 20 mg day^{-1} for adults, over 70 years. Uptake via ingestion accounted for over 90% of the total uptake (inhalation plus ingestion) into the body. It should be noted that the daily incremental uptake resulting from exposure to facility emissions was about three orders of magnitude *less* than the allowable daily intake for the various pollutants.

Evidence to support claims of adverse public health effects resulting from emissions to atmosphere is tenuous. In an examination of five case studies involving hazardous waste incineration that were often cited by environmental action groups as demonstrating adverse health effects such as cancers, birth defects, *etc.*, the following common features were noted.[38]

- Most of the reports relied on single newspaper articles, activist newsletters, overtly biased observers, or unreliable secondary sources. In four cases, no data were offered in support of the allegations.
- Research studies were quoted incompletely or out of context, often reversing the conclusions arrived at by the original workers. Typically, there had been no control group against which the allegations could be tested.
- Most of the allegations were based on self-reported symptoms uncorroborated by medical examination, and recall bias was inherent in all of the retrospective studies, particularly in well-publicized cases.
- The nature of the allegations hardly varied from facility to facility. A disproportionately small group of people generated the majority of the allegations, and continued to voice them long after these had been disproved or discredited.

These deficiencies in reporting have also been noted in well-publicized cases of environmental damage at Love Canal and Seveso.[39] In contrast, well designed

---

[38] R.C. Pleus and K.E. Kelly, 'Health Effects of Hazardous Waste Incineration—More of the Rest of the Story', Environmental Toxicology International, Seattle, 1994.

studies investigating allegations of ill health have tended to confirm the absence of statistically significant adverse effects relative to the control groups.[38-40] However, the propensity for some sites to cause nuisance cannot be denied, and it is possible that general loss of amenity coupled with experience of 'reversible' adverse health effects such as smarting eyes, temporary respiratory difficulty, headaches, *etc.*, is extrapolated to suggest the site as a causative agent of genetic and foetal damage or carcinogenesis when manifested in the exposed population. In the UK, careful analysis of health statistics by the Small Area Health Statistics Unit (SAHSU) has reversed a previous claim of a cluster of cases of cancer of the larynx in the vicinity of a chemical waste incinerator[41,42] and a study of ten other waste solvent and oil incinerators also found no evidence of an association with increased incidences of cancer of the lung or larynx.[42] Congenital malformations allegedly linked with a chemical waste incinerator in the Welsh district of Torfaen were examined in a study by the Welsh Office,[43] in which causation was not shown. A claim that an increased incidence of twinning among cattle and humans was observed in the vicinity of a chemical waste incinerator in Scotland[44] received much publicity, but careful analysis of the statistics showed a random occurrence of marginally raised rates over the entire study area, with no correlations between the operational and post-operational phases of the site nor with the handling or otherwise of specific chemicals assumed to have had oestrogenic properties.[45]

Emissions to atmosphere have been central to allegations of adverse health effects made against waste combustion plants. The atmospheric exposure pathway has also been implicated in operations such as landfills.[40] The atmosphere has the capacity to transport a wide range of chemicals, from volatile organics (as vapour) to nonvolatile organics and metals (as particulate matter, in dust emitted from the site). While the tenuous nature of the more serious allegations of adverse health effects attributed to waste management operations should be borne in mind, it is nonetheless important to appreciate the difficulties involved in carrying out health studies. Transport of chemicals away from the site via other environmental media (for example, groundwater or surface run-off) generally results in a more localized adverse effect for which causation is in theory easier to establish or disprove, since evidence of contamination is often retained in the transport medium and a limited and identifiable number of people come into contact with the contamination. Except in cases of sustained and continuous emissions, sampling of the atmosphere may not elicit a record of contamination,

[39] E. M. Whelan, 'Toxic Terror—The Truth behind the Cancer Scares', Prometheus Books, New York, 1993.
[40] G. M. Marsh and R. J. Caplan, in 'Health Effects from Hazardous Waste Sites', ed. J. B. Andelman and D. W. Underhill, Lewis Publishers, Michigan, 1987.
[41] A. C. Gatrell and A. A. Lovett, 'Burning Questions: Incineration of Wastes and Implications for Human Health', Research Report No. 8, North West Regional Research Laboratory, University of Lancaster, 1990.
[42] P. Elliot, M. Hills, J. Beresford, I. Kleinschmidt, D. Jolley, S. Pattenden, L. Rodrigues, A. Westlake, and G. Rose, *Lancet*, 1992, **339**, 854.
[43] Welsh Office, 'The Incidence of Congenital Malformations in Wales, with Particular Reference to the District of Torfaen, Gwent', Welsh Office, Cardiff, 1985.
[44] O. L. Lloyd, M. M. Lloyd, F. L. R. Williams, and A. Lawson, *Br. J. Ind. Med.*, 1988, **45**, 556.
[45] P. W. Jones, *Br. J. Ind. Med.*, 1989, **46**, 215.

and causation has to be deduced by indirect means. In addition, epidemiological studies have to take into account a number of features of exposure to waste sites.[3,40]

- The generally small size of the exposed population, resulting in loss of power in discriminating a site-specific effect from its background frequency.
- Mobility of the exposed population, introducing confounding factors relating to duration, type, intensity, and timing of exposure.
- The ubiquitous nature of the chemicals to which the population, including those in the vicinity of waste sites, is exposed. There are few chemicals unique to any particular site.
- Assessing the effects of exposure to mixtures of chemicals, and allowing for exposures unconnected with the site under investigation (for example through smoking and the workplace).

In summary, emissions to atmosphere from waste treatment and disposal facilities can and have caused loss of amenity and nuisance to surrounding populations. Claims of serious and irreversible health effects, such as congenital abnormalities and cancer, especially as a result of exposure to emissions representative of normal, day-to-day operations, are not substantiated by the evidence to date. However, this lack of evidence should not encourage an attitude of complacency within the waste management industry, since the robustness of the data for or against these claims is open to question and certainly can be improved. Given the difficulties inherent in any epidemiological study of small population groups, the availability of data relating to exposures experienced by the general population is vital in order to establish suitable controls. The role of General Practitioners in identifying health effects that may have environmental origins have been recognized by the British Medical Association,[46] while the activities of organizations such as SAHSU will improve the design and reliability of studies into adverse health effects.

6 Mitigation of Impacts

There are two forms of mitigation that can be applied to environmental contamination: cleanup of the contaminated medium together with isolation of the exposed population from the contamination, and implementation of control measures at source in order to reduce or terminate the emission. Mitigation of impacts caused or that could potentially be caused as a result of emissions to air is achieved through the latter route since air cannot be 'cleaned' in the sense that contaminated soil or water can, nor can the exposed population be isolated from the contamination, say by the erection of a fence or by being provided with an alternative source as in the case of water.

The most effective mitigation measures are preventative, *i.e.* a combination of prudent siting and the inclusion of appropriate abatement and control systems in the design of the plant. Siting a facility such that population exposure is minimized is the single most effective means of avoiding adverse health effects in humans as a result of direct exposure to emission. However, exposure via the

---

[46] British Medical Association, 'Hazardous Waste and Human Health', Oxford University Press, Oxford, 1991.

terrestrial foodchain is also of concern, since this pathway can account for a significant proportion of the total intake of a chemical. Exclusionary siting criteria aimed at minimizing the impact of emissions to atmosphere can include the following:

- not in an area prone to atmospheric inversions or similar unfavourable dispersion conditions;
- not in an area of prime agricultural value;
- not in a location upwind of a major conurbation; and
- not in the vicinity of other sensitive industrial users.

These should not be regarded as absolute criteria; rather, they are guidelines that, applied in an iterative manner to a range of potential sites along with other environmental and geographical criteria, serve to maximize protection of the environment irrespective of the precise technology or operation envisaged on the site.

Design measures would address both controlled and uncontrolled emissions. Incorporation of air pollution control equipment on combustion units is established practice across the whole of the waste management industry, and the proposed Directive from the European Union for standardization of emission limits irrespective of the type of waste incinerated will further tighten control over such units. Control measures aimed at minimizing fugitive release would include the enclosure of waste handling areas, extraction and scrubbing of air in the workplace prior to release to the surrounding environment, extraction and scrubbing of gases released during chemical reactions, prompt replacement of covers on drums after sampling of the contents, incorporating gas collection systems in landfills, covering emplaced waste in a landfill on a daily basis, *etc*. During the operational life of the plant, monitoring of emissions at source combined with monitoring of soil, grass, and air in the vicinity of the facility provides a means of management control that bridges the link between releases to atmosphere and the resulting impact on the environment.

Recycling Waste Materials: Opportunities and Barriers

J. L. GASCOIGNE AND S. M. OGILVIE

1 Introduction

In the late 1980s, the European Community developed a set of environmental objectives incorporated into the first amendment of the Treaty of Rome in July 1987 and from these a waste strategy[1] has been evolved with the following order of priorities:

- to prevent and reduce waste arisings at source;
- to increase recycling and re-use of materials and products; and
- to safely dispose of unavoidable wastes.

The pressure to improve waste management towards higher priority actions is increasing steadily. As one of its actions based on its waste strategy, the European Commission set about preparing a series of Directives with the intention to promote waste minimization and to set recycling targets for particular product groups and materials, for example, packaging, tyres, and batteries. The aims were to enact these during the 1990s. The implications are likely to have considerable impact on recycling practices in the UK.

To date, a batteries Directive[2] has been introduced. This Directive requires Member States to 'ensure the efficient organization of separate collection and, where appropriate, the setting up of a deposit system'. This applies to batteries containing certain dangerous substances but may only affect about 10% of batteries on the market. The packaging and packaging waste Directive[3] has been adopted by the Council following convening of the Conciliation Committee. This committee, comprising of representatives of both Council and Parliament, was set up to agree a joint text, following the Council being unprepared to accept one of the 19 amendments arising from the European Parliament's second reading. Member States must now make arrangements to implement this Directive within 12 months. This Directive introduces range targets for the recovery and recycling of packaging wastes (50–65% for recovery, including energy recovery, and

---

© UKAEA 1995.

[1] A community strategy for waste management: communication from Commission to Parliament SEC (89) 934 Final, Brussels, 18 September, 1989.
[2] 91/157/EEC, EC Directive on Batteries and Accumulators Containing Certain Dangerous Substances.
[3] 92/C263/01, Proposal for a Council Directive on Packaging and Packaging Waste, 12 October, 1992.

25–45% for recycling). The problems of certain types of other wastes are being tackled as priority waste streams initiatives.

The approach of 'Priority Waste Streams' was launched at the end of 1990 with the objective of bringing together representatives from government, industry, and environmental interests to quantify the fate of major waste streams, to analyse the options for waste minimization, re-use, recycling, and disposal, and to set targets for these, and then to obtain agreement from interested parties on how the responsibility for achieving the targets should be shared. After this negotiation stage, any necessary legislation would be proposed by Brussels. The intention of this approach was to cut down on the length of time taken to draft, negotiate, adopt, and implement Directives. This approach is currently beset with problems because of lack of resources at the Commission, resulting in concentration of efforts, for the time being, on just four priority waste streams. These are: (i) used tyres; (ii) chlorinated solvents; (iii) end-of-life vehicles; and (iv) clinical wastes. The Commission has asked for the four project working groups to conclude their work by early 1994 so that their success can be evaluated. More recently, it has been announced that there will be a formation of a Priority Waste Stream Group on End-of-Life Electrical and Electronic Equipment.

In the UK, the Environmental Protection Act (1990) has been introduced. Part II of this Act relates to 'wastes' and brings into force, among other measures, a Duty of Care on those involved with waste handling, a requirement for waste collection authorities to draw up recycling plans, and statutory obligations for the payment of 'recycling credits' to promote efforts to increase recycling wastes. The Government White Paper entitled 'The Common Inheritance' set a recycling target of 'half the recyclable fraction of household waste' (about 25% of this waste stream) by the year 2000. Furthermore, in its Second Year Report on the Environment White Paper, the Government has committed itself to using market-based mechanisms rather than regulation. Published in October 1992, this report states that 'in future, there will be a general presumption in favour of economic instruments. (Para. 3.46)'. The use of these so-called market-based instruments is being favoured for achieving desired waste management aims. For example, the idea of a levy on landfilling of wastes is being considered as a means of encouraging the market to adopt the higher priority options of the EC waste strategy by discouraging the reliance on cheap landfill disposal. The Department of the Environment (DoE) has published a report on the impact and feasibility of introducing a landfill levy[4] and a study on the 'externalities' of landfill and incineration has been published.[5]

More recently, the pressure to move to environmentally preferred options has increased with the announcement that the Department of Trade and Industry (DTI) and DoE were introducing a scheme of 'producer responsibility', initially directed at packaging waste, to develop a plan for recovering between 50% and 75% of the specified waste stream by the year 2000.[6] Senior representatives of

---

[4] Coopers and Lybrand, 'Landfill Costs and Prices: Correcting Possible Market Distortions', Department of the Environment, HMSO, London, February, 1993.

[5] CSERGE, Warren Spring Laboratory and EFTEC, 'Externalities from Landfill and Incineration', Department of the Environment, HMSO, London, November, 1993.

[6] 'Department of the Environment and Department of Trade and Industry Challenge the Packaging

retailers, fillers of packaging, and manufacturers of packaging and packaging materials had been invited to produce a plan by Christmas 1993 to meet the following objectives.

- Producers (retailers, fillers, and manufacturers of packaging and packaging materials) must take a share of the responsibility for what happens to packaging once it has served its original purpose, while minimizing the packaging deemed essential.
- Producers would show that they accept that responsibility by delivering targets for the amount of used packaging which they recover. Recovery targets of above 75% are unlikely to be practicable by the end of the decade. Therefore, producers are asked to commit themselves to ensuring that between 50% and 75% of all packaging waste (the precise level to be agreed with Government) is recovered by the year 2000.
- Producers would bear any extra costs of setting up systems to meet the recovery targets.
- Government will emphasize to producers the importance of their taking immediate action to ensure that a recycling infrastructure—collectors and processors—continues to be available to them, so that they can meet their recovery targets.

Plans had to be compatible with UK and EC competition law. If no industry commitment was shown or industry decided that legislative backing was essential, then a move towards a legislative approach to mandating producer responsibility would be necessary. The Producer Responsibility Industry Group (PRG) published its plan[7] in February 1994 for public consultation.

2 The UK Waste Context

Total UK waste arisings amount to just over 400 million tonnes y^{-1}. The Department of the Environment's Digest of Environmental Protection and Water Statistics[8] gives estimates for the annual waste arisings for the UK by sector. A summary is given in Table 1. This figure of 400 million tonnes y^{-1} should only be taken as an indication of the approximate order of magnitude of UK waste arisings because estimates in certain sectors are only quoted to the nearest 5 or 10 million tonnes. Research work continues with the aim of improving the reliability of these waste statistics.

In some sectors, good waste management practices exist. The level of recycling of wastes occurring is relatively high. For example, the amounts of demolition and construction wastes recycled have been estimated at 45%,[9] of which some 80% of asphalt road plannings are recycled. On the other hand, only some 5% of household waste is recycled currently. Generally, for techno-economic reasons,

---

Industry on Recycling', 27 July, 1993.
[7] Producer Responsibility Industry Group (PRG), 'Real Value from Packaging Waste', February, 1994.
[8] Digest of Environmental Protection and Waste Statistics, Department of the Environment, HMSO, London, No. 15, 1992.
[9] Arup Economics and Planning, 'Occurrence and Utilization of Mineral and Construction Wastes', Department of the Environment, August, 1991.

Table 1 Estimated total annual UK waste arisings

Sector	Annual arisings Mtonnes	Date of estimate	Status[c]
Agriculture	80	1991	NC
Mining and quarrying			
colliery and slate	51	1990	NC
china clay	27	1990	NC
quarrying	30	1989/90	NC
Sewage sludge	36	1991	PC
Dredged spoils	43	1991	PC
Household	20	—	C
Commercial	15	—	C
Demolition and construction	32	1990	C
Industrial			
blast furnace and steel slag	6	1990	C
power station ash	13	1990	C
other	50	—	C
Total	402	—	—

[c]NC = not classed as a controlled waste under the terms of the Environmental Protection Act (Controlled Waste Regulations) 1992; PC = sewage sludge is classed as a controlled waste as defined in the EPA (CWR) 1992 except when disposed of on agricultural land or within the curtilage of the sewage works at which it arises; dredged spoils are classed as a controlled waste when licensed for disposal under the Food and Environmental Protection Act. See Schedule 6 of the Collection and Disposal of Waste Regulations, 1988; C = controlled wastes under the terms of EPA (CWR) 1992.

recycling rates tend to be higher for the more homogeneous wastes: heterogeneous wastes are technically more difficult to deal with.

At Warren Spring Laboratory (now merged with AEA Technology to form the National Environmental Technology Centre), the Waste Research Unit has concentrated its monitoring and analysis expertise towards relatively heterogeneous wastes such as household wastes and commercial wastes. For example, the National Household Waste Analysis Project funded by the Department of the Environment continues presently. Its aims are:

- to provide data on the composition and weight of household waste which is representative of the UK as a whole;
- to have the potential to predict the composition and weight of waste arisings at a local (*e.g.* local government authority) level; and
- to provide information which will underpin work to identify more efficient resource and waste management practices in the UK, including waste minimization, design for recycling and disposal, waste collection practices, and treatment and residue disposal.

Resulting from this project, the typical composition of UK household (dustbin/garbage can) waste has been determined and is given in Figure 1.

The recyclable components of household waste are shown in Figure 2. The

Figure 1 Typical composition of UK household (dustbin) waste

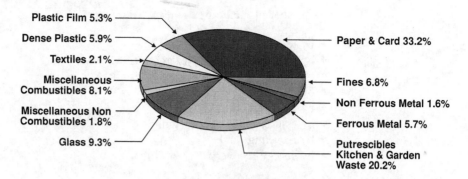

Figure 2 Recyclable components of household (dustbin) waste

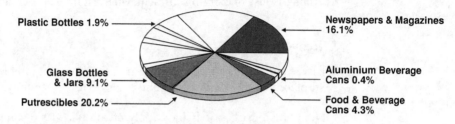

Table 2 Recyclable components of household waste

Figure 2 Recyclable Components of Household (Dustbin) Waste

Component	Weight/ %	Potential available amounts* /Ktonnes year^{-1}
Newspapers and magazines	16.1	2335
Plastic bottles	1.9	276
Glass bottles and jars	9.1	1320
Putrescibles	20.2	2929
Food and beverage cans	4.3	624
Aluminium beverage cans	0.4	58
Totals	52.0	7542

*Calculated using a basis of 14.5 million tonnes year^{-1} household waste arising

potential recovery of substantial amounts of usable materials is also clearly shown in Table 2.

When deciding and selecting measures within the broad framework of current policy and strategy, it is important to keep the following statements in mind.[10]

(i) Waste is material in the wrong form or place—it takes energy to put it back into productive use—*i.e.* energy is the ultimate resource.

(ii) Waste is a consequence of resource use. Of itself, it is not evidence of failure. Inefficient use of resources and poor waste management is.

Resource conservation is the key to reduction of wastes. This can be achieved

[10] J. R. Barton, 'Priorities for the Future', Paper W93014, Warren Spring Laboratory, to NSCA 1993 Workshop 'Waste Management, New Approaches, New Priorities', 23/24 March, 1993, Lincoln College, Oxford.

by a number of methods:

- by reducing consumption;
- by improving primary extraction processes;
- by fabrication/manufacturing design;
- by improving product life;
- by lightweighting;
- by re-use/refilling/recharging;
- by recycling; and
- by energy recovery.

Recycling is just one option for achieving resource conservation.

3 Recycling

In a free market situation, there are four basic, linked requirements which need to be in place before materials recycling can occur successfully. These are:

(i) there must be a reliable supply of suitable waste materials;
(ii) there must be the means to collect these materials and to transport them to a place where they can be re-processed;
(iii) there must be the means to re-process these materials into suitable raw materials and products; and
(iv) there must be available markets for the raw materials and products produced by the recycling process.

Economic considerations have a major influence on whether these four basic requirements can be achieved all at the same time. Failure to achieve one of the requirements will result in failure for materials recycling.

Additional factors influencing the uptake of recycling include efficiency of the recycling process, quality of the input material, quality of the recycled output materials, profit margin potential (to ensure a reasonable pay back period on investment in recycling equipment and technology), economic and environmentally appropriate outlet for wastes generated during the recycling process and location of recycling (*e.g.* close to major sources of waste arisings or close to customers for the recycled materials, *etc.*). The net effect of all these factors is that an optimum level of recycling will exist for a particular material which is dependent on a balance of these interacting factors. In the real world, a 100% recycling level is unlikely to be an optimal solution in waste management.

UK Materials Statistics

Recycling is not a new phenomenon. The process industries have practised forms of recycling and by-product utilization for many years. Economic reasons have been the driving force to do this. This is reflected somewhat in the UK materials statistics given in Tables 3 and 4—particularly for the metals industries where trends in scrap use indicate stability.

Table 3 UK materials statistics 1992

Materials	UK consumption/ Ktonnes	UK production/ Ktonnes	Consumption of scrap/ Ktonnes	Scrap use as % of consumption
Paper and board	9567	5128	3086	32
Container glass	1746	1742	459	26
Plastics	3491	2117	300	9
Aluminium	646	489	245	38
Ferrous	13 420	17 588	5297	39

Table 4 Trends in UK scrap use/Ktonnes year^{-1}

Materials	1984	1985	1986	1987	1988	1989	1990	1991	1992
Paper and board	2003	2067	2147	2310	2439	2601	2876	2954	3086
Container glass	147	214	—	245	255	310	372	385	459
Aluminium	206	182	166	167	200	220	201	195	245*
Ferrous	6688	5657	5324	5787	6111	6265	6397	5298	5297

*Change of methodology introduced.
Source: Digest of Environmental Protection and Water Statistics.

4 General Issues Affecting Recycling

Contamination

One of the main technical barriers to recycling is that of contamination. For example, household wastes are mixtures of potentially reclaimable materials and gross contamination. There will also be minor contamination by dirt, grease, moisture, and other materials. Many contaminants can be removed by efficient sorting, cleaning, and refining operations. Some contaminants are more difficult (sometimes impossible) to remove, particularly if they are chemically or physically bound into the structure of the materials.

There are two main categories of contaminant:

(i) those which are not removed during pre-treatment and processing operations and which impair the quality of the recycled material or product—commonly referred to as *residual contaminants*; and

(ii) those which can be removed by processing but where removal reduces the yield of the reclaimed product, extends processing times to allow contaminants to be reduced to acceptable limits, or leads to discharge of toxic fumes, effluents, or solid waste, which requires additional abatement measures—referred to as *non-residual contaminants*.

Table 5 summarizes the most common contaminants present in recyclable materials.

In metals, the presence of other metals can lead to microstructural defects which cause brittleness, poor surface finish, and cracking. For example, the residual contaminants, copper, tin, and nickel cause these defects in steel during hot rolling operations. The maximum allowable copper content for carbon steels varies between less than 0.15% and almost 0.5%, depending on the type of steel,

Table 5 Potential contaminants in recycled materials

Recycled material	Residual contaminants	Non-residual contaminants
Iron and steel	Copper, tin, nickel	Zinc
Aluminium	Iron, silicon	Lithium, glass, siliceous dirt, magnesium, zinc, tin, lead
Paper	Flexographic inks (>10%), water-resistant coatings	Adhesives, wire, staples, plastics
Glass	Iron and chromium colourants	Metals, ceramics
Plastics	Fillers, colourants	Other polymers, bacteria, inks, labels, adhesives
Compost	Heavy metals	Glass

the extent of hot working and cold forming operations, and the amounts of other impurities present, particularly tin. Zinc is present in steels as a non-residual contaminant in the form of electro-deposited coatings and zinc phosphate primers. Zinc is removed as emission of zinc fume. Although it is technically feasible to remove the zinc, the process requires longer refining times and creates additional waste disposal problems.

In the case of aluminium recycling, typical contaminants present in reclaimed scrap are shown in the table. The main problems are caused by alloys with other elements and other metals which are not removed before remelting of the scrap takes place. Can-stock may contain copper, magnesium, silicon, iron, manganese, and zinc; foil contains magnesium and silicon. Metal additions eventually reach an upper limit where properties of the alloy become adversely affected. Thermodynamic barriers prevent the removal of silicon, iron, and other metals which cause brittleness and loss of ductility, strength, and fracture toughness in the reclaimed metal. Elements which cannot be removed during remelting and refining processes are controlled in the final product by blending of the scrap with higher qualities of low residual scrap or pure aluminium metal. In contrast, lithium, magnesium, and zinc increase the rate of melt oxidation and therefore of dross formation during remelting and loss of useful alloying elements which can also lead to a reduction of mechanical properties. Tin, cadmium, and lead do not alloy with aluminium but, since their melting points are lower than for aluminium, they tend to be molten at the forming temperatures for aluminium products, leading to metallurgical defects.

Contaminants in wastepaper dictate the standard and the quality of the final product. Wastepaper is used as a substitute for primary pulp for writing, printing, wrappings, and tissues, and as bulk or packaging grades where the use of primary pulp is uneconomic. The most difficult contaminants are latex adhesives, plastics, and more recently flexographic inks, which are not removed by current flotation de-inking technologies, and water-resistant coatings which prevent or slow down the pulping process. The non-residual contaminants have the effect of reducing the quality of the final product; for example, from a pulp substitute to a packaging grade.

Apart from adventitious contaminants, wastepaper is degraded by reprocessing which reduces the length of the fibres and hence the mechanical strength of the product. The presence of these degraded fibres is essentially as a contaminant which reduces the quality of the resultant paper and limits its range of application. The removal of the short fibres constitutes part of the 'shrinkage' which occurs when paper is recycled.

For plastics, there is also a degradation of mechanical properties because of the high temperature moulding processes which reduce polymer chain lengths. Because these shortened polymers cannot be removed, various additives, for example, impact modifiers, are used to improve mechanical properties.

Plastics derived from household waste usually consist of a mixture of different polymer types, which cannot be processed easily unless they are separated. This fundamental difficulty of mixtures of different polymers means that identification and separation systems must be used. Compatibilizers are available to process some polymer combinations, but the products tend to be low value products such as wood substitutes. Plastics from household waste are also contaminated with dirt, labels, printing inks, and food residues which must be removed by cleaning processes. However, it is not usually possible to return recycled plastics to applications where good aesthetics are required; for example, transparent film. Pigmenting has to be used to mask the presence of contaminants. Apart from the aesthetic limitation, recycled plastics are generally not suitable to be used for food and beverage packaging applications because of the difficulties of ensuring completely reliable sterilization of bacterial contaminants and complete removal of chemical contaminants which may have diffused into the plastics. Chemically recycled plastics are an exception to this.

Fillers used to make plastics into composite materials are the most intractable contaminants, and the materials have little recycling potential apart from use as a fuel.

For glass, aesthetic appearance is the main property that is affected by the presence of contaminants. The colour of glass (*i.e.* green or amber) is dependent on the iron and chromium content and the chemical state of the metal ions in the glass. Different colours of glass cannot be reliably separated by automatic methods, and chemical methods for removing the metal ions from the glass are not available. As a result, colourless glass cannot be made from cullet contaminated with coloured glass, and although amber glass can be made from mixed amber and green cullet, the amount of green cullet acceptable in amber glass is limited. The net result is that most mixed cullet can only be used for green glass production, where the UK market is limited (15% of the total container glass market).

Metals are the contaminants of major concern in the applications of compost derived from household waste. Heavy metals such as lead, cadmium, and mercury originate from batteries, household dust, and various household chemicals, and can become more concentrated by the composting process. As few fundamental data are available on the uptake of heavy metals from composts, specifications of heavy metal concentrations for sewage sludge tend to be used as guidelines. Because no technology has been developed to remove heavy metals from compost, their presence as contaminants may limit the applications of composts to low value uses.

In summary, contaminants clearly present a barrier to recycling either in making the reclaimed materials unusable in a few cases or, more generally, to degrade properties and to limit the range of application to lower value products than for primary materials.

Collection

The typical composition data for the UK dustbin (US: garbage can) show that greater than 50% of the contents are 'recyclable' materials. It is estimated that the UK currently recycles about 5% of its household waste. The government has set a target of 25% recycling by the year 2000 (*i.e.* half of the recyclable waste). The practicalities of achieving this target need serious consideration.

Three major options for recovering recyclable materials exist.

(i) Bring systems, for example, bottle banks, paper skips, *etc.*
(ii) Collect systems, for example, kerbside collection, door-to-door, *etc.*
(iii) Centralized treatment.

Bring systems have been highly successful in the UK (especially for glass recycling). They are cheap to operate and recover the recyclable materials in a clean and relatively well segregated state. To maximize collection efficiency, it is believed that a high density of bank/skip sites is necessary, greater than one site per 2000 head of population. At such high levels of coverage, recoveries of up to 30% of the available material are possible.

Collect systems are viewed as the means to recover larger amounts of clean, well segregated recyclables. The reliance on the householder to deliver his recyclables to the 'recycling centre' is avoided, thereby ensuring greater participation in recycling. There are a number of kerbside schemes in operation in the UK; for example, the Sheffield Recycling City Project Kerbside Scheme, Milton Keynes, and Adur on the south coast of England, which have been set up to test the economic viability and practicality of such systems. Initial observations are as follows.

- The costs of operating such systems can be high (typically £100 tonne^{-1}) compared with conventional collection but that it may be possible to achieve commercial viability by judicious improvements in productivity; for example, co-collection of recyclables with normal refuse collections, split bin systems, planning of collection routines, *etc.*, careful re-tendering, and shrewd negotiations on prices for recovered materials.
- High participation levels are achievable; for example, the Milton Keynes scheme has claimed a participation rate* of 69% of all households in its first pilot area covered.

*Participation rate defined as the number of households setting out designated recyclables at least once in a 4-week period divided by the total number of households offered the collection during this 4-week period (expressed as a percentage).

- High recovery rates are possible; for example, the Sheffield kerbside scheme has collected up to 70% of the available plastic containers.

Kerbside collection could theoretically recover recyclables far in excess of the government 25% target, but only with the collection of putrescibles, and then at a cost! In practice, schemes without putrescibles collections have typically achieved 16–24% diversion rates* of dry recyclables.

In the USA, kerbside collection generally involves a combination of 'collect' and centralized treatment at Materials Reclamation Facilities (MRFs). This idea has also developed in the UK. Costs of the kerbside/MRF scheme can frequently be offset by the incorporation of commercial wastes, which tend to be easier and more profitable to recycle.

Municipal Solid Waste (MSW) centralized treatment systems have the advantage of access to **all** of the potentially recyclable material in household waste. The disadvantage is that the recyclables are grossly contaminated by this stage of the waste management process if collected as initially unsegregated wastes. Currently, the technology available for sorting, cleaning, and processing the recyclables can only recover material suitable for low grade re-use. Such recovered material commands lower prices than clean, well segregated materials.

Some countries have considered more draconian measures to ensure that maximum levels of recovery of materials are attained; for example, the German Packaging Ordinance/DSD system, where all packaging waste materials must be taken out of the public waste stream by whatever means suitable to industry (*e.g.* taking back packaging, private collection systems, or deposit schemes). This separate collection scheme for packaging waste relies on heavy subsidies which are provided from 'green dot' levies. This system has been so 'successful' in recovering packaging wastes that amounts collected for certain materials (notably plastics) have exceeded the home capacity to reprocess. The effect of this has been that recovered materials are being offered at prices (sometimes negative prices!) so low that German supplies of recyclable materials are seriously affecting the commercial viability of collection operations in other countries, including the UK. A 'level playing field' through introduction of a harmonizing EC packaging and packaging waste directive may solve this problem.

Generally, the main barriers to effective collection are economic. The main problem is to balance costs and the quality of the scrap that can be obtained. Technical development is required mainly to improve the economics.

Standards

Manufacturers demand that raw materials conform to specifications which strictly limit the nature and levels of contaminants tolerable. These specifications are often based on contaminant levels found in primary raw materials rather than 'adequate for purpose' levels. Thus it is necessary that recycled materials meet

---

*Diversion rate is defined as the total weight of materials recovered within a local government area (local authority) divided by the total household waste available for disposal within the local authority (expressed as a percentage).

these specifications in order to compete effectively. However, different end-uses can require different standards, so that reclaimed materials could complement primary raw materials in providing a range of products to suit all market requirements. Quite often, low contaminant levels can be achieved by dilution with purer materials.

Frequently, product specifications for raw materials quality are unnecessarily stringent for reasons related to perceived end-product quality requirements rather than 'fitness for purpose'. Quality means 'conformance to requirements' and as such, the terms 'high quality' and 'low quality' really have no place. (For example, in driving a car, if the requirement is solely to get from point A to point B, then a Mini is just as much a 'quality' car as a Rolls). Increased adoption of 'adequate for purpose' standards for raw materials specifications would remove a technical barrier to recycling and enable recycled materials to compete more effectively against virgin raw materials.

5 Technical Issues Affecting Recycling of Specific Materials

Glass

The UK container glass industry produced about 1 740 000 tonnes of glass containers in 1992 comprising: 230 000 tonnes amber glass (13.3%); 319 000 tonnes green glass (18.3%); and 1 192 000 tonnes colourless and other similar glass (68.4%). The actual amount of cullet collected for recycling was 459 076 tonnes (26.3% of total container glass production) of which the dominant fraction was green and mixed coloured.

The amount of waste glass which glass manufacturers can utilize in their operations depends on the desired colour of their products and the colour of the waste cullet available. Manufacturers have commented that too much poorly colour-segregated cullet is reclaimed (classed as mixed cullet even if it is predominantly colourless glass contaminated with a small amount of green glass). This mixed cullet is usually consigned to green glass production; green and amber glass container production capacity in the UK is much less than the capacity for colourless glass container production reflecting the relative UK demands for each colour, thus an imbalance exists between supplies of recovered glass and demands for each colour.

Colour segregation of waste glass is essential to achieve higher recycling rates by avoiding 'out of specification' cullet batch rejections. Nowadays, most glass manufacturers are tending to make payments for mixed waste glass **only** in situations where there has been a long-standing arrangement to do so. Mixed coloured glass collection is discouraged. Currently, some 92% of cullet collection is colour-separated.

Modern, properly equipped cullet benefication systems are designed to remove extraneous materials inherent to glass packaging. They will remove magnetic contaminants and non-magnetic metal (*e.g.* aluminium bottle caps and neck rings). Paper, plastic labels, cardboard, and similar items can be removed effectively. However, unexpected contaminants (*e.g.* brick, stones, ceramics, broken china, dirt, *etc.*) are not reliably removed. The technology to remove these contaminants effectively needs to be developed.

There are technical limits to the proportion of cullet which can be incorporated into the melt. Typically, the green melt can accommodate over 70% of green cullet whereas the colourless melt can be restricted to <20% colourless cullet in circumstances where containers destined for pressurized end-use are produced. Developments should concentrate on raising the level of cullet tolerated in the melt without detriment to the final product quality.

Effective collection and segregation can be achieved economically by the use of bottle banks. However, at best, bottle banks can recover about 30% of the total waste glass available. It is likely that source segregation schemes and kerbside schemes could achieve recovery levels around 70%. Therefore a sizeable fraction of the total glass available for recovery would still remain in the household waste stream and this fraction is highly likely to be grossly contaminated.

Glass collected from bottle banks and kerbside schemes can be recycled easily to make more glass containers because it is relatively clean and well-segregated (closed loop recycling). If the remaining grossly contaminated glass fraction occurring in the household waste can be recovered, then new markets for this material will be necessary unless the technology to sort and clean can be developed. Possible markets identified recently include composite roof covering materials, glass-fibre making, reflective beading, stone substitute in road making, and underground drainage tubes (a geotextile non-woven fabric sleeve filled with crushed glass).

Paper and Board

There are few technical barriers to paper recycling provided it is well separated into grades specified by the paper and board industry.

From the paper mill viewpoint, recycling waste-paper means a process of recycling fibres. Paper additives and fillers can represent a substantial proportion of the purchased weight of the waste-paper. When the waste-paper is processed, there is a reduction in length of the fibres as a result of the processing. Fibre fines are not retained on the web of the paper making machine. The loss of dirt, contraries, fillers, and fibre fines plus the possible excessive moisture content of the waste-paper is termed as 'shrinkage'. Some concerns have been raised about the possible occurrence of heavy metals (from the printing inks) and dioxins (from chlorine bleached paper) in the sludge from de-inking. Evaluation work is needed to establish the significance of such concerns. Lower shrinkage grades of waste-paper are favoured for better yields and reduced quantities of by-products to be disposed of.

Technically, it is possible to make products entirely from secondary fibre but it is not possible to utilize waste-paper to produce products requiring a higher quality of fibre than that contained in the waste itself without the quality of the final product being downgraded. However, the development of the technology to reduce loss of fibre fines to a minimum without loss of final paper quality would have great potential. For example, it has been claimed that the addition of 1% of chitin to pulp increases the strength of the paper, speeds up the rate at which water drains from the pulp, and increases the quantity of fibre retained when making sheets of paper, thus opening up opportunities to use cheaper, weaker fibres without losing quality.

Increasing the range of 'adequate for purpose' papers containing a minimum of 75% recycled fibre would lower barriers to recycling, but these papers would have to become readily accepted in the market place. Until now, specifications have been established for educational paper and computer listing paper.

Paper from household waste is usually graded as 'mixed' and tends to be suitable only for low grade board products, test liners, and industrial paper towels. A simplified classification of grades, understandable to the general public, could encourage better separation of paper types so that the higher grade uses of this waste could be achieved. Otherwise, improved mechanical sorting/separation processes which can recover higher grades economically would be needed. Currently, economics suggest that the waste-to-energy option for recovery of paper from mechanical separation plants is favoured.

Metals

Ferrous scrap is a relatively cheap source of iron for the Iron and Steel producing industry, but little if any ferrous scrap comprises 100% steel. There are other elements present such as:

- alloying elements originally incorporated to provide the desired physical and metallurgical properties;
- adventitious or contaminant elements;
- contaminants which have not been completely removed during processing of the raw ferrous scrap; and
- materials which are metallurgically bonded to the ferrous metal, *e.g.* plating, galvanizing, and brazing metals.

The principal technical barrier to ferrous metal recycling relates to the presence of non-ferrous metals and non-metallic materials which remain as contaminants in the scrap after processing. Some contaminants (non-persistent residuals) can be removed in the steel making process but usually entail additional processing costs; persistent residuals cannot be removed. Also there are costs of dealing with emissions, for example, zinc oxide from galvanized steel.

The technology improvements required should be directed towards (a) dealing with residuals effectively and (b) dealing with non-persistent residuals whilst maximizing yields and efficiency. The development of better methods of quantifying the amounts of non-ferrous metals in scrap could improve the productivity of scrap processing to give quality conforming steel products.

Technical barriers to ferrous and non-ferrous metals recycling are concerned with contamination by other metals and other non-metallics. Identification of contaminants and their separation from the scrap by effective methods could be helped enormously by proper 'design for recycling' of metal products and components.

Plastics

Plastics recycling is a relatively new activity compared with the more established scrap processing of metals, paper, and glass. Approximately 265 000 tonnes year^{-1} are currently recycled of which post-use recycling amounts to about 100 000 tonnes year^{-1}.

Low packing density is a major problem in the efficient collection of waste plastics. For example, a 30 m^3 container typically holds about 0.5 tonne of uncompacted plastic and a lot of air!

Once collected, plastic needs ideally to be sorted and separated by polymer type so that higher grades of reclaimed materials are produced which can compete against virgin raw materials for the manufacture of higher quality plastic products. The technology for optical sorting combined with X-ray fluorescence detection (for polyvinyl chloride) is being developed. These methods are currently limited by speed (about 3 containers s^{-1}) of sorting. Other sorting techniques are based on physical separation methods of granulated plastics. For example, density separation of polymer types using hydrocyclones or use of resistance to micronization to separate polyvinyl chloride (PVC) from polyethylene terephthalate (PET). Development of sorting techniques to commercial viability is continuing.

Where plastics are mixed to the extent that it is not possible to separate effectively, coprocessing of certain polymer combinations may be possible using compatibilizers. The end products tend currently to be low value articles such as plastic fence posts and the costs of the high value compatibilizer chemicals can sometimes make the end products uncompetitive. The development of modifiers which can improve mechanical properties of recycled materials could open up new higher value markets. Other technologies for processing mixed plastics are developing rapidly. For example, a sintering technique has produced a satisfactory rigid plastic sheet/board material.

Several new initiatives by motor manufacturers (BMW, Mercedes, and VW) and polymer producers (GE Plastics, BASF, and Hoechst) have been set up as pilot dismantling schemes for automotive plastic scrap. The general aims are to develop methods to re-use selected plastics (*e.g.* design of components to simplify dismantling and the reduction in the number of different polymer types used to aid re-processing and re-use). These initiatives take a long term view of plastics recycling from cars. In the short term, the technology to deal with the plastics already in circulation is needed.

Research has also been directed towards chemical techniques for recycling of plastics. Alcoholysis of PET to produce the raw materials for regenerating PET polymer appears to be promising, especially where use of any end product for food packaging might be considered. Pyrolysis of plastics to yield gases, liquids, and a solid residue has potential for fuel use. Similarly, hydrogenation of plastics is in the early development stages for the production of a refinable oil product. The variability of the waste plastic and economic problems have resulted in a failure to exploit this research so far.

Batteries

Although it is technically possible to recycle almost all the major constituents of the major small battery systems, the feasibility of wholesale battery recycling is currently limited by:

(i) the costs and difficulties associated with many of the post-use battery collection schemes currently favoured by municipal authorities;
(ii) a lack of cheap and efficient plants for sorting mixtures of batteries into concentrates suitable for existing recycling processes;
(iii) a lack of proven technology for recycling battery mixtures; and
(iv) the low intrinsic value of the most widely used battery systems (alkaline–manganese and zinc–carbon).

The expense and difficulty associated with collection and sorting combined with the cost of recycling currently restrict the extent to which small batteries are recycled.

Improvement in the economics of small battery recycling is needed otherwise other disposal options such as perpetual storage and landfill will be chosen for dealing with collected batteries. The debate continues as to the environmental significance of harmful emissions possible as a result of adopting such options.

Compostables

The reasons for the limited use in recent times of composting to treat amenable feedstocks, despite its potential benefits, are related to the diverse range of compostable wastes and their low perceived impact on urban areas; cheaper alternative dispersal routes (especially to land or landfill); a lack of technical information regarding the composting process, feedstock, and product characterization; and a poor perception of composting as a modern treatment option (largely related to poor product quality, inappropriate end-uses, and technical difficulties).

Possible contamination by toxic substances and heavy metals of composts produced from urban wastes represents a major barrier to their use in horticulture. Methods aimed at the exclusion of these contaminants at source or prior to composting are needed (*e.g.* collecting compostable waste separately from each household). For composts where concern over toxic metal contamination exists, alternative possible uses include use in construction as a fill or as a sound-proofing material, for example, sound reduction barriers along motorways. Contamination by glass and plastic is also a problem where further research and development into removal techniques for compost product improvement needs to be continued.

Textiles

Textile recycling has a long history in the UK, but it has been affected by the decline in the textile industry brought about by cheaper textile imports. Profit margins are tight and some operators have been forced to cease trading.

Textile reclamation is a very labour-intensive industry and therefore has high processing costs. Technology improvements need to be aimed towards reducing processing costs; for example, mechanical sorting by colour and grade, to improve profit margins in what has been a contracting industry in the UK. The key to increased textile recycling appears to be market development of new outlets for textile waste.

Waste Oils

Although the amounts of waste oil arisings and the fate of waste oils are not known with any accuracy, it is clear that oil wastes generated by DIY motorists (amateur car mechanics) represent the most significant contribution to oil lost to the environment. Provision of more reception facilities either at garages/filling stations or at civic amenity sites along with greater public awareness are necessary to improve collection rates. Oil wastes are perhaps a good example of where the challenge of 'producer responsibility' could be applied. There would be no barrier to selling the additional collected oil to existing users.

Waste oil is recycled in the following ways.

(i) *For use as a fuel* after removal of water, sludges, and the splitting of emulsions. Tight controls on emissions from combustion plants are becoming increasingly important factors affecting the uptake of this option.

(ii) *Oil-laundering and re-refining.* In the case of oil-laundering, the oil recovered mainly from the industrial lubricants category is blended with virgin oils after treatment to meet the standards for its original use or downgraded use. The re-refining process produces lube oil of adequate quality for re-use. Re-refining has ceased in the UK for economic reasons.

There are no significant technical barriers to waste oil recycling other than meeting emissions regulations when using waste oil as a fuel and the ability to achieve consistent quality levels of re-refined product from highly variable input materials.

6 Materials Case Study: Options for Scrap Tyre Recycling

Arisings and Composition

Although scrap tyres and other waste rubber present disposal problems, they are also a potentially valuable source of secondary raw materials and energy, because of their composition and the amounts which are produced.

Total used tyre arisings for 1989 were estimated as about 28 million, about 2.5 million of which were truck or bus tyres.[11] These figures have remained fairly steady for the last few years, because the numbers of new tyres sold has remained fairly constant. A car tyre weighs about 8 kg and a truck tyre about 45 kg. Hence the total weight of scrap tyres in 1989 was just over 300 000 tonnes. Current estimates of annual arisings are between 400 000 and 468 000 tonnes.[12–14] Of

---

[11] C.J. Burlace, 'Scrap Tyres and Recycling Opportunities', WSL Report LR 834, March, 1991.
[12] Elm Energy and Recycling (UK) Ltd. promotional literature.
[13] L. Oxlade, 'Tyred out: a burning issue', *Chem. Br.*, July, 1992, p. 582.
[14] 'Waste Tyres: Industry Faces Hard Choices', *Plast. Rubber Wkly.*, January, 1992, p. 9.

Table 6 Approximate proportions of the components in tyres

Component	Steel-braced radial car tyre/%	Radial truck tyre/%
Rubber compound	86	~85
Steel	10	~15
Textile	4	<0.5

Table 7 Typical composition of tyre rubber compound/wt%

Material	Proportion/wt%
Rubber hydrocarbon	51
Carbon black	26
Oil	13
Sulfur	1
Zinc oxide	2
Other chemicals*	7

*Includes inorganic fillers, organic vulcanization activators and accelerators, and processing aids.

these, about 50% are sent to landfill, about 30% for recovery and 20% for retreading.[13] Only about 9% of the arisings were used for energy recovery.

Rubber is used in a variety of other products apart from tyres, some of which are listed below:

Conveyor and elevator belting
Transmission belting
Hose and tubing
Coated fabrics and sheeting
Inflatable life-rafts
Carpet underlay
Cellular products
Footwear components
Seals
Hot water bottles

No figures are available for waste arisings from these products, but some idea of likely amounts of waste can be gained from production figures. In 1992, about 300 000 tonnes of rubber compound was used in non-tyre applications. The amount consumed is roughly equivalent to the amount discarded.[15]

The composition of a typical steel-braced radial car tyre and radial truck tyre are shown in Table 6.

Typical composition of tyre rubber compound is shown in Table 7.

Apart from being a source of materials, tyres are a potential source of energy, with an overall calorific value (CV) of about 31–32 MJ kg^{-1}, about 20% more than a typical UK coal. The polymers and oils in tyres can have CV values up to 42 MJ kg^{-1}.[11]

[15] International Rubber Study Group, personal communication, September, 1993.

Treatment and Disposal Options

The variety of treatment and disposal methods for scrap tyres can be categorized into final disposal routes and methods which involve recycling. Final disposal routes include landfill or incineration without energy recovery.

Recycling methods can be sub-divided into those which recover materials and those which recover energy. Materials recovery methods include retreading, granulation, rubber reclaim, and 'whole tyre' uses. Energy recovery methods include incineration, gasification, and pyrolysis, which can also be regarded as a materials recovery process.

Retreading. Some scrap tyres can be used again for their original purpose after retreading. In 1990, about 4 million retreaded car tyres and 1 million retreaded truck tyres were sold in the UK, giving market shares of about 18% and 34% in the respective sectors. About 35% of scrap car tyres are suitable for retreading, and an optimum market share for retreads to maintain a steady market has been calculated as 26%.[11]

The retreading process involves removing the surface from the tread section or the whole outer surface of the casing, and applying a new tread. The rubber removed is a good source of crumb for use in materials applications. There should not be any safety implications if the process is carried out correctly, but retreaded tyres have a poor reputation, despite having to comply with strict standards. One way of reducing the number of scrap tyres for disposal would be to promote retreading up to its optimum level.

Reclaim. Rubber reclaim is a method of devulcanizing rubber, so that it can be used to replace a certain amount of virgin rubber in new rubber products. Finely ground rubber is heated and treated with reclaiming chemicals, petroleum oils, and solvents under medium pressure. The product is a powder which is further plasticized by addition of fillers such as carbon black or clay. Recently improvements to the process have been made by minimizing energy use and eliminating the final refining process.[16]

Rubber reclaim activity declined in the UK in the late 1970s. Current activity is estimated at less than 50 000 tonnes annum^{-1}.[15] There has been a recent revival of interest in reclaim as a method of adding value to a low value, low cost material. However, reclaimed rubber has difficulty in competing with virgin rubber when prices are low. In addition, difficulty in achieving a consistent quality means that the product can only be used in down-market applications such as doormats.[15]

Granulation. Production of rubber granules, or crumb, is well-established in this country. There are several companies involved in granulation, and a number of suppliers of equipment, to produce granules of varying sizes. In 1991, UK rubber crumb production was estimated at 60 000 tonnes. About half of this was from scrap tyres, the remainder from other rubber scrap. About half of the tyre fraction was from buffings from the retreading industry.[11] In 1991 rubber crumb

---

[16] 'Recycling natural and synthetic rubbers', Paper to Rubberplas 84 conference, Singapore, 12/13 March, 1984.

was being imported because UK production fell short of demand. However, recent increases in the amount of crumb produced means that the market is now saturated unless new outlets can be found.

Uses for rubber crumb include brake linings, landscaping mulch, absorbent for oils, hazardous wastes, and chemical wastes, carpet backing, and sports surfaces. A recent example of the latter use is the jogging track built for President Clinton at the White House, which uses asphalt containing scrap tyres, and has a surface made from mulched rubber gaskets.[17]

In 1991 Rosehill Polymers of West Yorkshire were awarded a DTI grant to develop products from a combination of tyre and plastics waste. Possible applications, were flexible insulating materials, heavy duty packaging, flooring, playground surfaces, and patio tiles.

A growing use for rubber crumb in the USA is as an additive to asphalt in road-surfacing. The Intermodal Surface Transportation Efficiency Act 1991 requires states to use recycled rubber of other materials in road surfaces. By 1994, 5% of federally-funded roads paved with asphalt must contain recycled scrap tyre rubber, and this percentage increases to 20% by 1997.[18] This will open up a huge market for recycled rubber, and could be considered in the UK as a means of enlarging the market and reducing landfill of scrap tyres. Road surfaces incorporating rubber crumb have been found to last about twice as long as conventional roads.

Methods of treating the surface of rubber granules have been proposed, which make the rubber more compatible with virgin rubber. Examples are the 'Tirecycle' process and Air Products' treatment method. The Air Products' process treats the surface of rubber crumb with an oxidizing gas containing fluorine, which improves the adhesion and compatibility of the crumb with other materials. The Tirecycle process treats the rubber with a liquid polymer which promotes cross-linking with other rubber. So far this process has not been successfully developed commercially.

A further potential use for shredded tyres is in wastewater processing, where they have been used to replace plastic in packed bed reactors. This use is still under development, but it is claimed that the recycled rubber outperforms conventional materials and is cheaper.[19]

Uses for Textile Fibre. The textile fibre from tyres is usually regarded as waste for disposal. However, some uses have been found for it. For example, in cars the material could be used for exhaust system suspension straps, inserts to reduce wear between metal parts and in mud flaps. Other possible uses include roofing felt and 'erosion blankets' to stabilize roadside embankments.[20]

Energy Recovery. The calorific value of scrap tyres can be recovered by thermal treatment using one of three main methods: incineration, pyrolysis, and

---

[17] 'No Spare Tyre for the President', *Mater. Reclam. Wkly.*, 4 September, 1993, p. 12.
[18] US EPA Reusable News, EPA 530-N-92-005, Fall 1992.
[19] 'Tyres Used to Clean Food Processing Wastewater', *Waste Environ. Today (News J.)*, 1992, **5(8)**, 3.
[20] 'Uses of Non-rubber Wastes from Scrap Tyres', *Waste Manage. Today (News J.)*, 1989, **2(2)**, 10 and 'First UK Rubber Recycling Enterprise at Scunthorpe', 1991, **4(10)**, 10.

gasification. These methods are differentiated by the amount of air supplied during the process. Apart from energy recovery, they also reduce the amount of solid waste for disposal, and can also produce usable materials.

Emissions are controlled under Part 1 of the Environmental Protection Act 1990, and the Prescribed Processes and Substances Regulations 1991. Operations must be authorized by Her Majesty's Inspectorate of Pollution (HMIP), who have issued Guidance Notes setting limits for emissions. In addition, operators must prevent, minimize, or render harmless all emissions, and must monitor both the operating conditions and the emissions from their plant.

Incineration. Thermal treatment of tyres with excess air is known as incineration. Several tyre incinerators are in existence, and a large new facility opened at Wolverhampton in November 1993. Advantages of incineration are that it is a well-established technology which generates useful heat and allows the recovery of steel and zinc oxide. However, combustion is difficult to control because of the high rate of release of volatile materials from tyres. Combustion must therefore necessarily be a two stage process.

Concern is often expressed about the emissions which result from tyre incineration, based on those produced in tyre dump fires. The clean-up equipment required to deal with these emissions includes filters or electrostatic precipitators for collecting particulates, and scrubbers to trap the sulfur dioxide and other gases. Clean-up costs are therefore high.

Scrap tyres have been used as fuel in cement kilns, and this use is popular in countries such as Japan, France, and Germany.

Pyrolysis. Thermal treatment in the absence of air is known as pyrolysis. Pyrolysis technologies have been developed using a number of different types of furnace, from rotary kiln and microwave to a thermolysis process using an oil bath. The solid, liquid, and gaseous products of pyrolysis all have potential uses. The oily liquid product can be used as a transportable fuel or as a chemical feedstock, while the gas can be used for process heating or to produce methanol. The solid products are carbon which can be used in new tyres or as a filter medium, and steel which can be used in the scrap metals industry.

By altering the conditions the products obtained can be controlled, and there are no large volumes of waste gases requiring expensive clean-up. Tyres lend themselves to pyrolysis because they are predominantly hydrocarbon, and so produce relatively consistent products. Problems have been encountered in scaling systems up to commercial operation, but pyrolysis continues to attract attention. Recent schemes include the AEA-Beven batch units which reached commercial production in 1992 and the microwave furnace of BRC Environmental Ltd. Crucial to the success of pyrolysis schemes is the existence of stable markets for the products.

Gasification. Thermal treatment in sub-stoichiometric quantities of air is called gasification. The product gas can be used directly in a gas turbine, which is more efficient than a steam turbine. In addition, the amounts of waste gas produced are smaller than for incineration, thus incurring less clean-up costs. As yet there are no commercial gasification plants operating in the UK, although plants exist in the USA, Brazil, and other European countries notably Sweden. These are not specifically designed to handle scrap tyres.

Whole-tyre Uses. Because of their resistance to degradation, and their resilient properties, use of whole or part tyres has been proposed for a number of applications. These include artificial reefs for coastal protection, fenders on boats and docksides, temporary roads, sound-proofing barriers on roads, and injection wells in landfill sites.

Landfill. Landfill of whole tyres is permitted in law, but in practice is discouraged. The DoE Waste Management Paper No. 26 recommends that whole tyres should not be landfilled, and that landfill of shredded tyres should be restricted to 5% of the landfill volume. The EC Priority Waste Stream discussions on tyres have recommended that landfill of tyres should decrease to near zero. The cost of landfilling tyres has risen to over £35 tonne^{-1} and figures as high as £80 tonne^{-1} have been quoted as typical charges for disposal in specialist dumps.[21] This is the principal incentive in looking for alternative disposal routes.

Trends and Markets for Materials

Incineration. The Royal Commission on Environmental Pollution (RCEP) Report on Incineration, published last year, concluded that incineration of waste should play an important part in the UK Government's waste management strategy. Scrap tyres have been accepted as a renewable energy source for the Non-Fossil Fuel Obligation (NFFO), which has recently been extended beyond its previous 1998 deadline. The Elm Energy tyre incinerator at Wolverhampton, which was opened in November 1993, will use about 21% of UK scrap tyre arisings, and generate 25 MW of electricity. Elm plan to open at least one more plant in the UK, at East Kilbride in Scotland. It is likely, therefore, that incineration will play an increasing role as a disposal route for scrap tyres.

Markets for Materials. Because of the large numbers produced, and the rise in the cost of landfill, scrap tyres have come to be regarded as a cheap source of material, which with the minimum of processing can be turned into a valuable product.

However, in view of the Elm Energy project and other incinerators, and the proliferation of proposals for small granulation or pyrolysis schemes, there is the real possibility that in the long term, scrap tyres might be in short supply. On the other hand, the market for granules is limited, and so there may not be outlets for all the crumb produced. Legislation similar to that in the US, which requires a minimum recycled content in road surfacing, would open up a huge new market for granulated rubber. Until then, potential entrants into the industry should ensure that their supplies and markets are guaranteed.

EC Discussions. The EC Working Group on the used tyres priority waste stream has now produced an Action Plan, which will be presented to the Commission. The plan proposes measures on prevention, collection, retreading,

---

[21] F. Pearce, 'Scrap Tyres: A Burning Issue', *N. Sci.*, 20 November, 1993, pp. 13–14.

recovery, and disposal. Among the targets are a decrease to near zero in landfill, a decrease of 10% in scrap tyre arisings, an increase in recovery to 60%, and an increase in retreading to 30%. It is intended to implement the Action Plan by means of a Recommendation and a Code of Practice, rather than a legally binding Directive. In the future, therefore, there should be increased support for development of alternatives to landfill for disposal of scrap tyres.

Concluding Remarks on Options for Scrap Tyres. Scrap tyres are a large scale source of secondary raw materials and energy, which at present pose problems of disposal. It is therefore reasonable to seek alternative methods of treatment and disposal. There are a number of these available, some of which recover material, others energy. The most developed in this country are retreading, use of rubber crumb, and incineration with energy recovery. Others with potential are pyrolysis, gasification, and 'whole-tyre' uses.

Alternatives to landfill of tyres will be encouraged following EC discussion, and it is likely that incineration will play an increasingly significant part in tyre disposal in the future, provided sufficient investment is made, and planning and environmental controls can be complied with. The existence of stable long-term markets is crucial for the success of schemes which recover material values from tyres.

7 Conclusions

Recycled materials will always have to compete with virgin raw materials on cost, availability, and quality. Many of the technical issues raised in this paper have highlighted the need to recover materials in a satisfactory state capable of use in preference to exploiting more non-renewable resources. These issues are inextricably linked to economics and geography. For increased recycling, techno-economic solutions will be required. Technological advances in materials handling and processing can help to reduce the costs of recycling materials with the desired quality, but market forces will dictate whether the economics are favourable for the use of recycled materials.

The concept of quality or conformance to requirements of materials is fundamental in recycling. *Technology* has a major part to play in the achievement of recycled material quality suitable for end use. The removal of contamination, the ability to collect recyclables efficiently, and extension of standards designated as 'adequate for purpose' are seen as the main areas for concentration of technical efforts.

Disposal of Nuclear Fuel Waste

K. W. DORMUTH, P. A. GILLESPIE, AND S. H. WHITAKER

1 Introduction

As of the end of 1991, nuclear generating stations supplied 17% of the electricity generated in the world.[1] The International Atomic Energy Agency estimates that by the end of 1992 about 135 000 Mg (heavy metal) of used nuclear fuel had been produced.[2]

Used fuel is not necessarily a waste, because it contains plutonium and uranium which could be used to produce more energy by recycling them back into a reactor. First, however, the used fuel would have to be reprocessed to separate these useful materials from the unwanted ones. Some 25% to 30% of the used fuel produced world-wide is currently expected to be reprocessed.[1] Reprocessing of used fuel is being done at facilities in France, India, Japan, Russia, and the United Kingdom.[3]

If used fuel is not to be reprocessed, it must eventually be disposed of. We will discuss the disposal of used fuel with emphasis on the Canadian disposal concept. There is no reprocessing of fuel in Canada at present and no definite plans for reprocessing. If the fuel is reprocessed in the future, we expect that a solid waste form such as borosilicate glass would be manufactured to incorporate most of the unwanted radioactive material and would be disposed of in much the same way as we will describe for used fuel.

Characteristics of Used Fuel

When used fuel is removed from a reactor, it is highly radioactive because of the decay of unstable atoms. It emits energy in the form of radiation. Because many of the unstable atoms have short half-lives, the activity decreases rapidly. However, used fuel will remain radioactive for a long time, because some of its unstable atoms have long half-lives. Much of the radiation is absorbed by the used fuel

---

[1] B. A. Semenov, 'Disposal of Spent Fuel and High-Level Radioactive Waste: Building International Consensus', *IAEA Bulletin*, 1992, **34**(3), pp. 2–6.
[2] F. Takats, A. Grigoriev, and I. G. Ritchie, 'Management of Spent Fuel from Power and Research Reactors: International Status and Trends', *IAEA Bulletin*, 1993, **35**(3), pp. 18–22.
[3] P. M. Chantoin and J. Finucane, 'Plutonium as an Energy Source: Quantifying the Commercial Picture', *IAEA Bulletin*, 1993, **35**(3), pp. 38–43.

Figure 1 A typical CANDU fuel bundle

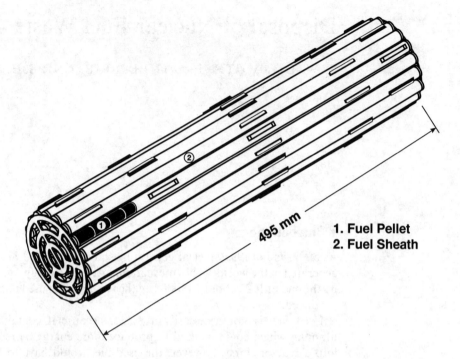

1. Fuel Pellet
2. Fuel Sheath

itself, generating heat within the fuel. As the activity decreases, so does the heat generated.

A single fuel bundle used in a CANDU™ reactor contains about 19 kg of natural uranium and is about 50 cm long and 10 cm in diameter, about the size of a fireplace log (Figure 1). It generates about one million kilowatt hours of electricity. Thus the value generated per unit volume of waste produced is very high, which allows elaborate measures to be taken to ensure its safe disposal. By contrast, 400 Mg of coal would have to be burned to produce the same amount of electricity as one fuel bundle.

Thus three characteristics of used fuel are particularly important when considering how it is managed: it is radioactive, it generates heat, and its volume per unit of electricity generated is small.

Potential Effects of Radiation

The radioactive elements in used fuel can be hazardous if swallowed or inhaled. Some also emit radiation that can penetrate into or through the human body from an external source. The effect depends on the dose received and on the time over which it is received. The radiation dose to humans is given in a unit called the sievert (Sv). The dose rate is commonly expressed in millisieverts per year (mSv y^{-1}).

We are all constantly exposed to background radiation, mostly from naturally occurring sources. Throughout the world, values of dose rate from natural background radiation range from about 1.5 to 6 mSv y^{-1}, with an average of 2.4 mSv y^{-1}.[4] In addition, we are exposed to radiation caused by human activities. Typical doses are 0.1 mSv from a chest X-ray and 0.03 mSv from a dental X-ray.[4]

It is not certain that low levels of radiation, such as background levels, cause any harmful effects. It is difficult to determine whether variations in disease rates are associated with the variation of natural background radiation with location, and increases in the frequency of cancer have not been documented in populations in areas of high natural background radiation.[5] Thus the effects of low doses and low dose rates cannot be determined directly. Instead, the probability of health effects, such as fatal cancer or serious genetic effects, is estimated using observations of effects from much higher doses and dose rates. The International Commission on Radiological Protection (ICRP) has estimated the probability of a radiation dose inducing a cancer in an adult of the general population to be $0.05\ \text{Sv}^{-1}$.[6]

Current Management of Used Fuel

Used fuel is currently stored in water-filled pools (wet storage) or in metal or concrete structures filled with inert gas or air (dry storage). Storage may be either at the nuclear generating station (as is the case at present in Canada) or at centralized storage facilities serving several generating stations.

Used fuel in storage is isolated to prevent the radioactive elements from contaminating the natural environment where they could be swallowed or inhaled. It is shielded by materials such as water or concrete to protect against external radiation. In addition, used fuel is cooled to remove the heat produced by radioactive decay.

In Canada, storage facilities for used fuel have been safely operated for over 45 years. They are licensed and inspected by Canada's nuclear regulatory agency, the Atomic Energy Control Board.

The Need for Disposal

Current storage practices, while safe, require continuing institutional controls such as security measures, monitoring, and maintenance. Institutional controls are not considered to be reliable for the indefinite future. Therefore, Canada, like other nuclear power producing nations, has seen a need to develop a method of disposal which would not require institutional controls to maintain safety in the long-term. This does not mean that society would not implement long-term institutional controls, but rather that if such controls should fail, human health and the natural environment would still be protected.

Disposal is intended to minimize any burden placed on future generations resulting from the nuclear fuel waste produced by the present generation, taking social and economic factors into account. Since the present generation derives a

---

[4] United Nations Scientific Committee on the Effects of Atomic Radiation, 'Sources and Effects of Ionizing Radiation', Report to the General Assembly, with annexes, United Nations, New York, 1988.

[5] National Academy of Science Advisory Committee on the Biological Effects of Ionizing Radiation, 'Health Effects of Exposure to Low Levels of Ionizing Radiation', National Academy Press, Washington, DC, 1990, pp. 383–395.

[6] International Commission on Radiological Protection, '1990 Recommendations of the International Commission on Radiological Protection', ICRP Publication 60, Annals of the ICRP, 1991, **21**(1–3).

significant benefit from the electricity, it ought to assume, to the extent possible, the responsibilities associated with disposal.

Alternatives for the Disposal of Nuclear Fuel Waste

Three types of disposal have been considered internationally by researchers investigating alternatives for the disposal of nuclear fuel waste:

(i) removal of the waste from the earth by transporting it into space;
(ii) transmutation, which would entail changing some of the radioactive elements in the waste to different elements, by nuclear methods, in order to reduce the long-term radiotoxicity of the waste; and
(iii) geological disposal, which would entail isolating the waste in a geological medium (an ice sheet, sediment or rock beneath the deep seabed, or sediment or rock on land) in such a way that maintenance and administrative controls would not be required in the long-term.

After more than 20 years of research and evaluation of alternatives, there is international consensus that geological disposal using a system of natural and engineered barriers is the preferred method,[1] and most countries with large nuclear power programs are planning for land-based geological disposal in repositories excavated at depths of several hundred metres. International research on land-based geological disposal of nuclear fuel waste has concentrated on five disposal media: plutonic rock (often called crystalline rock), salt, clay (or shale), tuff, and basalt. In each country, the decision to focus on a particular type or types of rock is made on the basis of the geological conditions within that country and a variety of other factors.

2 The Canadian Disposal Concept

General Requirements for a Disposal Concept

The Canadian research and development on disposal has focused on developing a concept that meets the following general requirements:

(i) human health and the natural environment must be protected;
(ii) the burden placed on future generations must be minimized, social and economic factors being taken into account;
(iii) there must be scope for public involvement during all stages of concept implementation; and
(iv) the disposal concept must be appropriate for Canada, that is, compatible with the geographical features and economic factors.

These general requirements were based on directives from the governments of Canada and Ontario, the regulatory documents of the Canadian Atomic Energy Control Board, the objectives for disposal established by the International Atomic Energy Agency and the Nuclear Energy Agency of the Organization for Economic Co-operation and Development, and the results of a public involvement program conducted in Canada.[7]

Features of the Disposal Concept

The proposed disposal concept is a method for geological disposal of nuclear fuel waste. Multiple barriers would protect humans and the natural environment from the contaminants in the waste. These barriers would be the container; the waste form; the buffer, backfill, and other repository seals; and the geosphere (the rock, any sediments overlying the rock below the water table, and the groundwater). Institutional controls would not be required to maintain safety in the long-term.

The waste form would be either used CANDU fuel or, if the used fuel were reprocessed in the future, the solidified high-level waste from reprocessing.[8] The low solubility of used CANDU fuel under expected disposal conditions would make it effective for retaining contaminants; thus it is an excellent waste form in its current state. The liquid radioactive waste that would result if used fuel were reprocessed would not be suitable for direct disposal, but such waste could be solidified to produce an excellent waste form, as is done currently in France.

The waste form would be sealed in a long-lived container to facilitate handling of the waste and to isolate it from its surroundings.[8] The container material and other aspects of the container design would be determined when developing a design for a potential disposal site.

The containers of waste would be emplaced in a repository excavated nominally 500 to 1000 m below the surface in plutonic rock of the Canadian Shield. Plutonic rock, such as granite, is formed deep in the earth by crystallization of magma and/or by chemical alteration. Plutonic rock of the Canadian Shield has characteristics considered to be technically favourable in a disposal medium, and it offers the greatest scope for siting in Canada because of its wide geographic distribution. The rock would protect the waste form, container, and repository seals from natural disruptions and human intrusion; would maintain conditions in the repository that would be favourable for long-term waste isolation; and would retard the movement of any contaminants released from the repository.

The repository would be designed to accommodate the subsurface conditions at the disposal site. It would be a network of horizontal tunnels and disposal rooms excavated deep in the rock, with shafts extending from the surface to the tunnels.[9] The greater the depth, the greater the minimum distance for contaminants to move from the disposal rooms to the surface, and the lower the likelihood of any natural disruption or inadvertent human intrusion. However, the temperature and stresses in the rock and the cost of construction and operation tend to increase with depth. The nominal disposal depth of 500 to 1000 m was chosen to strike a reasonable balance among these considerations.

---

[7] M. A. Greber, E. R. Frech, and J. A. R. Hillier, 'The Disposal of Canada's Nuclear Fuel Waste: Public Involvement and Social Aspects', Atomic Energy of Canada Limited Report, AECL-10712, COG-93-2, 1994.

[8] L. H. Johnson, J. C. Tait, D. W. Shoesmith, J. L. Crosthwaite, and M. N. Gray, 'The Disposal of Canada's Nuclear Fuel Waste: Engineered Barriers Alternatives', Atomic Energy of Canada Limited Report, AECL-10718, COG-93-8, 1994.

[9] G. R. Simmons and P. Baumgartner, 'The Disposal of Canada's Nuclear Fuel Waste: Engineering for a Disposal Facility', Atomic Energy of Canada Limited Report, AECL-10715, COG-93-5, 1994.

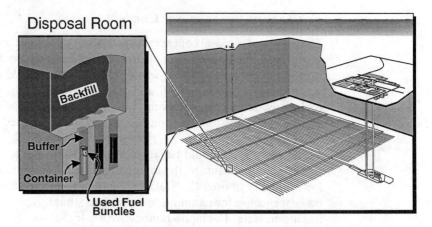

Figure 2 Illustrative disposal facility showing borehole emplacement

The containers of waste would be emplaced either in the rooms or in boreholes drilled from the rooms[9] (Figure 2). Each container would be surrounded by a buffer, a material that probably would contain clay.[8] A buffer around the container would limit the rate of corrosion of the container, limit the rate of dissolution of the waste form should groundwater seep into the container, and retard the movement of any contaminants released from the waste form and the container.

Each room would be sealed with backfill and other repository seals, made of materials containing clay or cement.[8] These seals would fill the space in the room; keep the buffer and containers securely in place; and retard the movement of any contaminants released from the waste form, container, and buffer.

All tunnels, shafts, and exploration boreholes would ultimately be sealed in such a way that the repository would be passively safe; that is, long-term safety would not depend on institutional controls.[8] The repository seals would keep people away from the waste and would retard the movement of any contaminants released from the disposal rooms. Research, both in Canada and internationally, has focused on clay-based and cement-based sealing materials.

The Canadian research and development on disposal has been generic in the sense that no site on which to implement the concept has been designated or sought. In fact, government policy is that no disposal site selection will be undertaken until the concept has undergone a public review and has been found acceptable by the government. This public review of the concept is now being conducted under the federal Environmental Assessment and Review Process. Atomic Energy of Canada Limited, a federal Crown corporation, is the proponent in this review and has prepared an environmental impact statement on the concept for disposal of Canada's nuclear fuel waste.[10]

On the basis of research and development on the disposal concept, we expect that the multiple barriers would protect human health and the natural environment far into the future.

[10] Atomic Energy of Canada Limited, 'Environmental Impact Statement on the Concept for Disposal of Canada's Nuclear Fuel Waste', Atomic Energy of Canada Limited Report, AECL-10711, COG-93-1, 1994.

3 Qualitative Discussion of Long-term Performance

The overall purpose of the repository would be to prevent radioactive contaminants in the waste from reaching the biosphere in amounts that would exceed criteria, guidelines, or standards. In evaluating the performance of a repository, it is common practice to make conservative assumptions regarding the sequence of events that could occur to cause release of material. This is not an attempt to predict what will happen, but an analysis of what could happen to cause release. To this end we consider the following potential sequence of events:

(i) groundwater present in the host rock contacts the containers;
(ii) the containers corrode;
(iii) groundwater contacts the waste form and contaminants are released from the waste form;
(iv) contaminants move through the repository seals;
(v) contaminants move through the geosphere; and
(vi) contaminants move through the biosphere where they can affect humans and non-human biota.

In this section, we qualitatively discuss the processes involved in this sequence of events and indicate how a repository is expected to perform in the long-term. In Section 4, we give a quantitative analysis for a particular choice of disposal system characteristics (a case study) and include there a discussion of potential natural and anthropogenic disturbances that could affect the performance of the repository.

Groundwater Movement

Throughout most of the Canadian Shield, groundwater saturates the rock and sediment to very near the surface. Water moves through fractures in the rock under the influence of gravity. Even rock that has no visible fractures has minute interconnected pore spaces through which water can move, albeit very slowly.

On topographic highs and surrounding slopes, groundwater tends to move downward from the water table (recharge). In topographic lows, which are often occupied by swamps, lakes, and streams, groundwater tends to move upward to the surface, where it discharges. The generally low relief of the Canadian Shield landscape provides a generally low driving force for groundwater movement. Thus the movement tends to be extremely slow.

Excavation of a repository would cause the rock adjacent to the tunnels and rooms to become unsaturated as groundwater drained from the rock into the excavation. The water that entered the excavation would be pumped to the surface. When a portion of the repository was sealed, groundwater would move from the saturated portion of the surrounding rock toward the repository until the rock, buffer, and backfill became saturated.

The time required for the more permeable parts of the rock to resaturate probably would be a few months to a few years, whereas it could take thousands of years for very low-permeability rock to resaturate. Because the buffer and backfill would have very low permeability, it would take several years to thousands of years for them to become saturated with groundwater, depending on the hydraulic properties of the surrounding rock as well as the geometry and permeability of the buffer, backfill, and other repository seals.

Container Performance

As a clay-based buffer absorbed water it would swell and exert pressure on the containers. Additional pressure would be exerted by the groundwater once the buffer was saturated. The containers would be designed to withstand these pressures.

Eventually, water would contact the containers, and, to be conservative in evaluating performance, it is assumed that this would occur immediately upon closure of the repository. The container material would be chosen to resist corrosion at the temperatures and under the chemical conditions expected in a repository. Research shows that both copper and titanium would corrode very slowly under such conditions.

Non-oxidizing conditions would limit the rates of container corrosion and waste-form dissolution. At field research areas on the Canadian Shield, the groundwater at potential disposal depths does not contain significant amounts of dissolved oxygen. This is because the rock contains large amounts of iron-bearing minerals that take up oxygen in the water coming from the surface. In a repository, the backfill would also contain large amounts of iron-bearing minerals that would deplete the water of free oxygen. Thus, in a repository, the groundwater is not expected to act as a long-term source of oxidants.

Within 500 years, the activity of the radioactive waste will be 5 orders of magnitude lower than when it came out of the reactor. The remaining activity will be caused primarily by the long-lived radioactive elements, such as iodine-129, caesium-135, technetium-99, plutonium-239, chlorine-36, and carbon-14. Thus the container would be designed to last at least 500 years to ensure that the waste would be completely isolated during the operation of the disposal facility and until there is a substantial decrease in the activity and heat output of the waste. Research into the corrosion of container materials and the structural strength of containers indicates that containers could be designed so that most would last for at least tens of thousands of years under the conditions expected in a repository.[8]

Despite thorough inspection, a few containers might have undetected manufacturing defects that could cause them to fail prematurely. On the basis of current manufacturing data, less than 1 container in every 1000 is expected to have a defect that could lead to an early failure.

Contaminant Release from Used Fuel

Should a container fail, groundwater would gradually enter the container through the penetrations. This groundwater in the container would eventually penetrate the corrosion-resistant fuel sheath and reach the uranium dioxide pellets. The majority of the new radioactive elements formed in the reactor are bound within the pellets. Because the pellets would dissolve extremely slowly under the conditions expected in a repository, a large proportion of the inventory of many of the radioactive elements would decay while still retained in the fuel pellets (an example of such an element is plutonium-239). A small proportion of the inventory of some radioactive elements is not bound within the pellets and would be released more quickly (these elements include iodine-129, caesium-135, technetium-99, and carbon-14). Over a period of 100 000 years, less than 1% of the mass of the radioactive elements formed in the reactor would be released. Thus the used fuel would be an excellent barrier even if there were no containers.

Any contaminants released from the used fuel would move through the groundwater in the container mainly by diffusion. They would be released through the small penetrations in the container wall into the groundwater in the buffer.

Contaminant Movement Through the Repository Seals

The low permeability of the buffer would inhibit groundwater movement; thus any contaminant movement through the groundwater in the buffer would be mainly by diffusion. As contaminants diffused through the groundwater in the buffer, they could be removed from the groundwater, either temporarily or permanently, through chemical reactions with minerals in the buffer and with the groundwater. This would tend to retard the movement of contaminants. Some radioactive elements, such as plutonium-239 and caesium-135, would be greatly retarded, whereas others, such as iodine-129, technetium-99, and carbon-14, would be more mobile.

Many of the radioactive elements that entered the buffer would not be released from the buffer because (i) contaminant movement by diffusion is slow, (ii) chemical reactions would tend to retard contaminant movement, and (iii) radioactive elements decay with time. The buffer would provide backup protection for radioactive elements expected to be retained in the waste form while they decayed. Material released from the buffer would enter the backfill or the rock.

Contaminants entering the backfill from the buffer would move very slowly through the backfill, because of its relatively low permeability and high porosity. Chemical reactions with minerals in the backfill and with the groundwater would retard contaminant movement to a greater extent than in an equivalent volume of buffer, because the pore volume of the backfill would be greater. Thus the backfill would retain some intermediate-lived radioactive elements, such as technetium-99, until little of them remained because of radioactive decay. The backfill would provide backup protection from radioactive elements expected to be retained in the waste form and the buffer. Material released from the backfill would enter the groundwater in the pore spaces and fractures in the rock.

Contaminant Movement Through the Geosphere

In moving groundwater, contaminants would move through the rock with the groundwater and would be dispersed. Even if the groundwater was not moving, contaminants would move through it by diffusion. In either case, chemical reactions with minerals in the rock and with the groundwater would tend to retard the movement of contaminants through the geosphere.

Plutonic rock bodies contain roughly planar zones of intensely fractured rock. These fracture zones are usually only a few metres thick, although some are regional faults that are tens of kilometres in length and tens to hundreds of metres thick. Fracture zones can be interconnected and can have relatively high permeability. Where they are permeable, fracture zones could be pathways for the relatively rapid movement of contaminants with the groundwater. A repository would be designed such that containers of waste would be separated from fracture zones by low permeability rock. Below depths of 200–500 m at field research areas on the Canadian Shield, blocks of rock have been identified that have very low interconnected porosity and permeability. These blocks lie between the fracture zones and are large enough to contain significant quantities of nuclear fuel waste. Contaminant movement in such rock would tend to be by diffusion and thus would tend to be very slow. A few metres of such rock would be an effective barrier to the movement of water and contaminants.

Contaminant Movement Through the Biosphere

The biosphere of the Canadian Shield consists of rocky outcrops; bottom lands with pockets of soil, marshes, bogs, and lakes; and uplands with meadows, bush, and forests. Contaminants could move through the biosphere via water wells, surface water, lake sediments, wetlands, soils, the atmosphere, and food chains. In moving through the biosphere, contaminants could be delayed, dispersed, and diluted, thereby mitigating effects on humans and other biota. On the other hand, there could be accumulation of material in some parts of the biosphere, which could serve to enhance these effects.

4 Quantitative Case Study of Long-term Performance

On the basis of the qualitative considerations of the previous section, it would be expected that the disposal system would be safe in the long-term. In this section, we will discuss a quantitative assessment of long-term performance.

The Canadian Atomic Energy Control Board requires that the predicted radiological risk from a disposal facility shall not exceed 10^{-6} for the 10 000 years following closure.[11] The radiological risk is defined as the probability that an individual of the 'critical group' will incur a fatal cancer or serious genetic effect due to exposure to radiation. This critical group is a hypothetical group of people assumed to live at a time and place and in such a way that its risk from the repository is likely to be the greatest.

To make quantitative estimates of the risk associated with disposal, it is necessary to be specific about the characteristics of the disposal system. Because no disposal site can be selected until after the concept is accepted, a case study was carried out by assessing a hypothetical disposal system having characteristics

[11] Atomic Energy Control Board, 'Regulatory Objectives, Requirements and Guidelines for the Disposal of Radioactive Wastes–Long-term Aspects', Atomic Energy Control Board Regulatory Document R-104, Ottawa, 5 June, 1987.

based on information derived from extensive laboratory and field research.[12] Many of the assumptions made tend to overestimate adverse effects. In addition to the sequence of events discussed qualitatively above, disruptions that could occur during the 10 000-year period were considered in the assessment.

Hypothetical Disposal System

The characteristics of the hypothetical repository were based on an engineering conceptual design for a repository in granitic rock at a depth of 500 m.[9] In the design, the waste form is used fuel. The disposal container has a titanium shell 6.35 mm thick and holds 72 used-fuel bundles. A packed particulate fills the void spaces in the container and supports the container shell against the external pressure. Each container is emplaced in a borehole drilled into the floor of a disposal room and is separated from the rock by a clay-based buffer material. The rooms, tunnels, and shafts are sealed with clay-based backfill and other repository seals.

Geological information for the case study was taken from the Whiteshell Research Area, and the geological setting of the repository is consistent with observations made there. The rock immediately surrounding the disposal rooms at 500 m depth has very low permeability, whereas the rock closer to the surface is more permeable. A highly permeable fracture zone is assumed to extend from the surface past the repository (Figure 3). No container is located less than about 50 m from the fracture zone.

For the case study, the critical group is a hypothetical rural household, located in the immediate groundwater discharge area associated with the hypothetical repository. It is assumed to be totally self-sufficient, deriving all of its food, water, fuel, and building materials from local sources. The environment of the critical group was derived using information from the Canadian Shield in Ontario.

Assessment Approach

The features, events, and processes that could have an influence on the risk to the critical group were considered. Teams of researchers identified and analysed them to determine which needed to be included when estimating the risk. For example, analysis showed that a strike by a meteorite large enough to disrupt the hypothetical repository has such a low probability that it need not be included. On the other hand, the critical group's use of a well located near the repository was considered sufficiently probable that it was included. Similarly, inadvertent intrusion into the repository by exploratory drilling in the future was considered sufficiently probable that it was included.

[12] B. W. Goodwin, D. B. McConnell, T. H. Andres, W. C. Hajas, D. M. LeNeveu, T. W. Melnyk, G. R. Sherman, M. E. Stephens, J. G. Szekely, P. C. Bera, C. M. Cosgrove, K. D. Dougan, S. B. Keeling, C. I. Kitson, B. C. Kummen, S. E. Oliver, K. Witzke, L. Wojciechowski, and A. G. Wikjord, 'The Disposal of Canada's Nuclear Fuel Waste: Postclosure Assessment of a Reference System', Atomic Energy of Canada Limited Report, AECL-10717, COG-93-7, 1994.

Figure 3 Repository and fracture zone assumed for the postclosure assessment case study

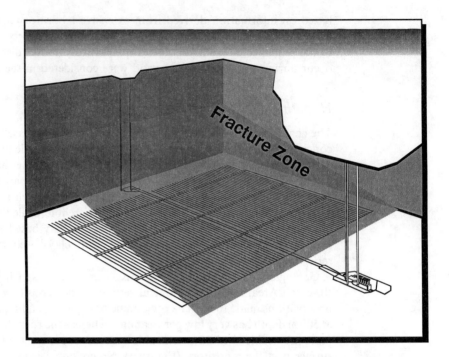

Mathematical models of the disposal system were constructed.[13–15] The models are based on knowledge gained through research and development on disposal, both in Canada and in other countries. The models estimate the magnitude of the effects as well as the probability that the effects would occur. The models take into account uncertainty in input data values, and they are used to analyse the sensitivity of the estimates to variations in the data.

In addition to estimating the radiological risk to an individual of the critical group, these models were used to estimate the concentrations in the biosphere of contaminants from the hypothetical repository. They were also used to estimate effects on four generic species (plant, mammal, bird, and fish) that live in the same area as the critical group.

Results of the Case Study

For all times up to 10 000 years, the estimated mean dose rate to an individual of

[13] L. H. Johnson, D. M. LeNeveu, D. W. Shoesmith, D. W. Oscarson, M. N. Gray, R. J. Lemire, and N. C. Garisto, 'The Disposal of Canada's Nuclear Fuel Waste: The Vault Model for Postclosure Assessment', Atomic Energy of Canada Limited Report, AECL-10714, COG-93-4, 1994.

[14] C. C. Davison, T. Chan, A. Brown, M. Gascoyne, D. C. Kamineni, G. S. Lodha, T. W. Melnyk, B. W. Nakka, P. A. O'Connor, D. U. Ophori, N. W. Scheier, N. M. Soonawala, F. W. Stanchell, D. R. Stevenson, G. A. Thorne, S. H. Whitaker, T. T. Vandergraaf, and P. Vilks, 'The Disposal of Canada's Nuclear Fuel Waste: The Geosphere Model for Postclosure Assessment', Atomic Energy of Canada Limited Report, AECL-10719, COG-93-9, 1994.

[15] P. A. Davis, R. Zach, M. E. Stephens, B. D. Amiro, G. A. Bird, J. A. K. Reid, M. I. Sheppard, S. C. Sheppard, and M. Stephenson, 'The Disposal of Canada's Nuclear Fuel Waste: The Biosphere Model, BIOTRAC, for Postclosure Assessment', Atomic Energy of Canada Limited Report, AECL-10720, COG-93-10, 1993.

Disposal of Nuclear Fuel Waste

Table 1 Percentage of a contaminant released by a barrier over 100 000 years

Contaminant	Half-life (years)	Barriers in the case study				
		Used fuel	Titanium container	Clay-based buffer and backfill	Low-permeability rock	All barriers combined
Strontium-90	29	0.05	≪0.001	1	≪0.001	≪0.001
Argon-39	270	8	0.08	≪0.001	≪0.001	≪0.001
Carbon-14	5730	6	60	0.8	0.007	≪0.001
Plutonium-239	24 000	≪0.001	100	≪0.001	≪0.001	≪0.001
Technetium-99	210 000	6	100	≪0.001	0.1	≪0.001
Iodine-129	16 000 000	6	100	10	5	0.03

the critical group, excluding the consequences of human intrusion by exploratory drilling, is 6 orders of magnitude lower than the dose rate associated with the radiological risk criterion specified by the Atomic Energy Control Board and 8 orders of magnitude lower than the dose rate from natural background radiation.

In estimating the risk associated with inadvertent human intrusion into the repository by exploratory drilling, it was assumed that institutional controls would gradually become ineffective over a period of 500 years. For all times up to 10 000 years, the estimated risk from inadvertent human intrusion by exploratory drilling is at least 3 orders of magnitude lower than the radiological risk criterion specified by the Atomic Energy Control Board.

The estimated concentrations of contaminants in water, soil, and air are so low that there would be no significant chemical toxicity effects from any contaminants potentially released from the hypothetical repository.

The dose to plants and animals is given in a unit called the gray (Gy). The dose rate is commonly expressed in milligray per year (mGy y^{-1}). For a wide variety of plants and animals, the dose rate from background radiation is 1 to 100 mGy y^{-1}, depending on factors such as diet, habitat, and way of life. For the case study, the dose rates to a plant, a mammal, a bird, and a fish were estimated to be lower than those from background radiation, which are themselves lower than dose rates known to cause harm.

Table 1 shows the potential effectiveness of the used fuel, the titanium container, the buffer/backfill, and the low-permeability rock surrounding the disposal rooms. For each of these barriers, the table gives an estimate of the percentage of a contaminant released by a barrier over 100 000 years. It also gives an estimate of the effect of all barriers combined. The smaller the percentage released, the more effective the barrier. Some implications of the data given in the table are discussed below.

The potential effectiveness of a barrier depends on the contaminant being considered. For example, the used fuel is a very effective barrier for plutonium-239, because it is bound within the uranium dioxide fuel pellets and is released only as the pellets slowly dissolve. The used fuel releases very much less than 0.001% of the mass of plutonium-239 over 100 000 years. Larger percentages of contaminants such as argon-39 and iodine-129 are released.

More than one barrier may be effective in limiting the release of a particular contaminant. For example, there are three very effective barriers for plutonium-239: the used fuel, the buffer/backfill, and the low-permeability rock. Each of these barriers releases very much less than 0.001% of the mass of this contaminant over 100 000 years. Similarly, there are two very effective barriers for strontium-90 and argon-39. For carbon-14, technetium-99, and iodine-129, there are three barriers that are either effective (release no more than 10% of the mass of these contaminants over 100 000 years) or very effective: the used fuel, the buffer/backfill, and the low-permeability rock.

The container is an effective barrier only for radioactive contaminants with half-lives that are short in comparison with the container lifetime. The examples given in the table are strontium-90 and argon-39. A longer-lasting container would show greater effectiveness for a material having a long half-life.

The percentage of a contaminant that would pass all these barriers in 100 000 years ranges from very much less than 0.001% to about 0.03%. Thus the combination of barriers is very effective for all the contaminants. The percentage that would pass all barriers in 10 000 years is very much lower than the releases over 100 000 years.

At times beyond 10 000 years, the repository is expected to have effects similar to those of a uranium ore body containing the same amount of uranium.[16] Thus in the event of glaciation, earthquakes, or a meteorite impact, the risk from a repository is expected to be about the same as the risk from a rich uranium ore deposit.

Thus the case study indicates that risks from a disposal system having the characteristics of the hypothetical system would meet regulatory requirements for the 10 000-year period with a large margin of safety. For times beyond 10 000 years, the risk is not expected to exceed that from a rich uranium ore deposit.

5 Implementation of Disposal

The case study employed many conservative assumptions. In particular, the waste was placed close to a fracture zone and the critical group used a water supply well drawing water from the fracture zone at the center of the contaminant plume from the repository. At any actual future disposal site, the waste would be located so as to allow the characteristics of the groundwater flow systems, the groundwater chemistry, and the rock structure to enhance the safety of the disposal system.[17] Therefore, a larger margin of safety than indicated by the case study would be expected.

A technically suitable site would be one at which it could be demonstrated that nuclear fuel waste disposal would meet all applicable criteria, guidelines, and standards for protecting human health and the natural environment. We believe that technically suitable disposal sites are likely to exist in Canada because

(i) the qualitative considerations of the performance of a repository indicate that a disposal system would be safe in the long-term;

[16] K. Mehta, G. R. Sherman, and S. G. King, 'Potential Health Hazard of Nuclear Fuel Waste and Uranium Ore', Atomic Energy of Canada Limited Report, AECL-8407, 1991.
[17] C. C. Davison, A. Brown, R. A. Everitt, M. Gascoyne, E. T. Kozak, G. S. Lodha, C. D. Martin, N. M. Soonawala, D. R. Stevenson, G. A. Thorne, and S. H. Whitaker, 'The Disposal of Canada's Nuclear Fuel Waste: Site Screening and Site Evaluation Technology', Atomic Energy of Canada Limited Report, AECL-10713, COG-93-3, 1994.

(ii) the numerical results of the postclosure assessment case study show a very large margin of safety between estimated effects and the regulatory limit, even though many conservative assumptions were made;
(iii) the geological conditions assumed for the case study are based on conditions at an actual location in plutonic rock of the Canadian Shield and are not unusual; and
(iv) plutonic rock is abundant on the Canadian Shield.

Disposal of nuclear fuel waste would proceed in sequential stages: siting (possibly ~ 20 years), construction (~ 5 years), operation (at least ~ 20 years and possibly more than 80 years to dispose of 5 to 10 million fuel bundles), decommissioning (~ 10 years), and closure (~ 2 years). Decommissioning and closure could be delayed to permit extended monitoring. The stages and activities of concept implementation are briefly described below.

The siting stage would involve identifying a site for a repository through community participation and technical investigation. From the large regions of plutonic rock available on the Canadian Shield, a small number of areas would be identified. These areas would be investigated in detail, a preferred site would be identified, and approval would be sought for construction of a disposal facility at that site.

The construction stage would involve constructing the facilities and systems needed to begin disposing of nuclear fuel waste. These would include transportation facilities and equipment, access routes, utilities, surface facilities, shafts, tunnels, underground facilities, and some or possibly all of the disposal rooms. All systems would undergo testing in preparation for full operation in accordance with legislative requirements.

The operation stage would involve transporting nuclear fuel waste to the repository, putting the waste into long-lasting containers, and emplacing the containers and sealing materials in the repository. At the same time, construction of disposal rooms would continue if all rooms had not been completed during the construction stage. The operation stage could begin with a demonstration of operation, during which the disposal rate or repository size might be limited. The construction schedule might also be affected.

After the operation stage, decommissioning would be delayed to allow for extended monitoring if the implementing organization, the regulatory agencies, or the host community required additional data on the performance of the filled, partially sealed repository. Similarly, after the decommissioning stage, closure would be delayed to allow for extended monitoring if the implementing organization, the regulatory agencies, or the host community required additional data on the performance of the sealed repository.

The decommissioning stage would involve the decontamination, dismantling, and removal of the surface and subsurface facilities. It would also involve the sealing of the tunnels, service areas, and shafts, and of the exploration boreholes drilled from them. The site would be rehabilitated and markers could be installed to indicate the location of the repository. Access to any instruments retained for extended monitoring would continue to be controlled.

The closure stage would involve the removal of monitoring instruments from any exploration boreholes that could, if left unsealed, compromise the safety of the disposal system, and then the sealing of those boreholes. During the closure stage, the objective would be to return the site to a state such that safety would not depend on institutional controls.

Activities such as obtaining approvals, characterization, monitoring, design, assessment of environmental effects, and management of environmental effects would be ongoing. Implementation of the disposal concept would entail a series of decisions about whether and how to proceed,[18] and the involvement of potentially affected communities would be sought and encouraged throughout all stages.[7]

A preclosure assessment case study of a hypothetical disposal system indicates that a disposal facility and transportation system similar to those assessed could be implemented while protecting the public, the workers, and the natural environment.[19]

6 Conclusions

Although current storage practice is a safe interim measure for the management of used fuel, eventually nuclear fuel waste must be disposed of. Research and development spanning more than 15 years indicates that the concept of disposal in plutonic rock of the Canadian Shield meets the general requirements for an acceptable disposal concept, and that implementation of this concept represents a means by which Canada can safely dispose of its nuclear fuel waste.

Acknowledgment

The Canadian research and development described in this article was performed as part of the Nuclear Fuel Waste Management Program, which is funded jointly by Atomic Energy of Canada Limited and Ontario Hydro under the auspices of the CANDU Owners Group.

---

[18] C. Allan, 'Building Confidence in Deep Geological Disposal of Nuclear Fuel Waste: Canada's Approach', Presented to the International Nuclear Congress INC93, October, 1993.

[19] L. Grondin, K. Johansen, W.C. Cheng, M. Fearn-Duffy, C.R. Frost, T.F. Kempe, J. Lockhart-Grace, M. Paez-Victor, H.E. Reid, S.B. Russell, C.H. Ulster, J.E. Villagran, and M. Zeya, 'The Disposal of Canada's Nuclear Fuel Waste: Preclosure Assessment of a Conceptual System', Ontario Hydro Report N-03784-940010 (UFMED), COG-93-6, 1994.

The Economics of Waste Management

D. W. PEARCE AND I. BRISSON

1 The Waste Disposal Problem

Disposal of solid waste is a major problem in both developed and developing countries. The problems are two-fold: environmental impacts, including health hazards, and costs of disposal. Environmental impacts arise from pollution associated with incineration, landfill, and recycling of waste. Health hazards arise from some air pollutants, from waste not disposed of to controlled outlets, from poorly managed waste sites, and from possible ground water contamination by leachate from landfill sites. Health hazards are, by and large, relevant to developing countries where disposal practice is often primitive.[1] The cost impacts are often overlooked. In high income countries (those with incomes over $8356 capita^{-1} in 1992, as classified by the World Bank) disposal costs may amount to some $30–60 billion annually. Medium income countries ($676–8355 capita^{-1} incomes) may have costs of some $5–12 billion, and low income countries (below $676 capita^{-1}) may have costs of $1 billion. The costs are uncertain because of the paucity of data for many countries, but these cost estimates amount to a fairly consistent 0.2–0.5% of GNP (Gross National Product).[1] Costs of this magnitude represent serious financial burdens in any country. In the developing world, waste disposal costs can easily dominate local government finance in a context where local taxation and borrowing is often difficult to secure.

Estimates of total waste arisings world-wide are few and far between. Table 1 provides some approximate estimates.

The problems of environmental, health, and cost impacts justify a focus on the economics of waste management. Oddly, despite these issues, the literature on the economic management of waste is sparse. In this paper we therefore outline a consistent economic theory of waste management, illustrating it with data from various case studies.

2 The Economic Approach to Waste Management

To understand the economist's approach to waste management issues it is necessary to explain some basic concepts. The basic philosophy is based on a

[1] D. W. Pearce and R. K. Turner, 'Economics and Solid Waste Management in the Developing World', Paper read to University of Birmingham conference—'Whose Environment?—New Directions in Solid Waste Management', May 1994.

Table 1 Municipal waste arisings and disposal world wide by income category

	Low income	Middle income	High income
Arisings capita^{-1}	200 kg annum^{-1}	300 kg annum^{-1}	600 kg annum^{-1}
Waste (\$ income)$^{-1}$	0.51 kg \$$^{-1}$	0.15 kg \$$^{-1}$	0.03 kg \$$^{-1}$

Source: Adapted from data in S. Cointreau-Levine, 'Private Sector Participation in Municipal Solid Waste Services in Developing Countries', Infrastructure and Urban Development Department, World Bank, Washington, DC, 1992.

utilitarian approach of balancing costs and benefits. Thus, no one waste disposal method is intrinsically 'good'. Rather, each method has to be appraised for its costs and its benefits. Cost and benefit in this context are measured according to impacts on human well-being or welfare (or utility). In turn, well-being and welfare are measured by what individuals are willing to pay (WTP) in the market place for benefits, or what they are willing to accept (WTA) by way of compensation for any costs. In essence, then, a benefit reflects an individual's preference for something; a cost reflects a 'dispreference' for something. So long as an economy is fairly freely functioning, prices in the market place will be a good measure of these preferences, since they are determined by the interaction of scarcity (supply) and willingness to pay (demand). Clearly, this relationship breaks down if the goods or commodity in question is not bought and sold in the market place. Environmental assets, such as clean air, clean water, biological diversity, and so on, are generally not bought and sold in this way. They are 'non-market' goods. However, they are also the subject of human preferences, so that if it is possible to find out what individuals *would* be willing to pay *if* there were markets, then the link between human preference and price can be re-established. The resulting prices are known as 'non-market prices' or 'shadow prices'.

In practice there are many ways of finding out WTP in non-market contexts. As an example, house prices often reflect the WTP for amenity, quiet, and reduced pollution: houses with these characteristics attract higher prices than those without them. There is, therefore, a 'surrogate' market in the environmental good and the statistical procedures used to find the implicit value of the good are part of the 'hedonic property price approach' to uncover WTP. In other cases, WTP can be elicited by direct questioning of respondents, just as a market research company determines the demand for new and untried goods before they are launched on the market. This direct questioning approach is known as 'contingent valuation'.

The economic approach is not, therefore, solely concerned with marketed goods and services. It extends quite explicitly to non-market contexts. Nor is economics just concerned with money. Money happens to be a convenient measuring rod for WTP which in turn is a measure of human preferences. The economic approach is quite explicitly anthropocentric—but this need not be a rejection of 'moral' views concerning the 'rights' of non-humans, or concepts of obligation, justice, and fairness. Often these concepts will be embodied in the WTP of individuals. Where they are not, however, the economic approach is still very powerful since it explicitly acknowledges the limited resources available for solving any problem. Approaches based on 'rights' often fail to acknowledge the budget constraint.

While freely functioning markets produce prices reflecting WTP and scarcity, what they do not do is capture any 'third party' effects. For example, the purchaser of a product from the supermarket does not pay directly for the cost of disposing of the packaging on the product. He or she may be paying for the supermarket's own waste since disposal of that waste attracts a commercial charge which the supermarket must bear and which is therefore likely to be incorporated in the price of the product. But the purchaser does not pay for his or her own waste disposal except indirectly through local or national taxation. There is, therefore, a third party effect, or 'externality', which the market system tends to ignore. In the same way, methane and carbon dioxide emissions from landfill sites are externalities: they are not accounted for in the charges for using landfill, but they do damage to others. Much of the economic approach is therefore dedicated to ways of ensuring that these externalities are incorporated in the cost of products and disposal options, *i.e.* to 'internalizing the externality'. In this way economic agents bear the full 'social' cost of their actions and the market mechanism will then be modified accordingly.

3 Integrated Waste Management: An Economic Theory Approach

While much is made of 'integrated waste management' (IWM) in policy discussions on waste, rational approaches to integration are few and far between. The economic approach offers a coherent structure for IWM. IWM can be approached in a sequential way.

The first question is how much waste should be produced. Put another way, what is the right balance between the costs of reducing waste at source and the benefits of doing so? In much of the environmental literature source reduction—the lowering of the ratio of waste to product—appears as a costless activity which is therefore ranked as a priority over all other means of disposal, recycling, *etc.* But source reduction is not costless. Consider food packaging. Apart from its contribution to consumer convenience, it prevents some foodborne disease and substantially reduces loss rates between food production and consumption. In the OECD countries perhaps as little as 2–5% of food output is lost between the point of production and the point of consumption. In the developing world that figure can easily rise to 50%. In the former Soviet Union, for example, some 20–30% of cereals production never reaches the consumer, in part due to inefficient organization of transport but also because of poor protection in transit. Some 50% of vegetables may not reach consumers. In Brazil it is estimated that 20% of rice output, 25% of flour output, and 30% of beans are lost during warehousing, transport, and sales. Certainly, packaging is only part of the solution to this problem, but it is a significant part.

The first stage in IWM, then, is some judgement balancing of the costs and benefits of source reduction. It is unlikely to be a detailed quantitative exercise, but even the practice of 'cost–benefit thinking' represents a major advance on much that passes for policy analysis.

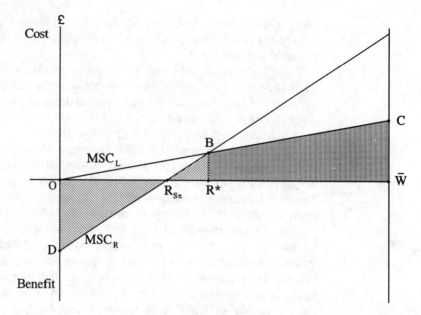

Figure 1 The basic analysis of final disposal *versus* recycling ($\bar{W}$ is the amount of waste, MSC_R is the 'marginal social cost of recycling', MSC_L is the marginal social cost of landfill)

The second stage of IWM involves analysis of the costs and benefits of the various disposal options, where we include recycling and re-use as a disposal option, with landfill and incineration as 'final' disposal. Again, the proper procedure is to weigh up the relevant costs and benefits of each option. Failure to do this produces 'environmentally correct' but analytically unsound policies such as making recycling a priority over landfill and incineration when this may not be warranted. Recycling is not a costless activity and costs, in turn, are surrogates for resources. If recycling has financial costs in excess of benefits, then this is a first warning signal that it may also not be economically sound in the sense of weighing up overall costs and benefits. Much the same applies to incineration which is discriminated against in much legislation without any obvious logic.

To illustrate the analysis required, Figure 1 sets out an elementary comparison of recycling and disposal where it is assumed that there is only one final disposal option, say landfill. More real world complexity can be introduced as the analysis proceeds.

The amount of waste is shown on the horizontal axis. Total waste arisings are $O\bar{W}$, and these are assumed to be 'optimal' in the sense that the source reduction decision has already been evaluated. The issue is how to distribute these waste arisings between recycling and landfill.

The vertical axis shows cost in money terms. Positive and negative costs are shown, but note that cost is shown above the point of origin and benefits below. Now consider the line MSC_R. This is the 'marginal social cost of recycling'. What this curve shows is the cost of recycling inclusive of any environmental impacts of recycling *less* any revenues obtained from selling the recycled materials. 'Social' simply serves to remind us that we are not just interested in the costs for the recycler but also any costs borne by third parties, *e.g.* because of pollution from, say, a de-inking process. 'Marginal' here simply means 'extra', so that the line shows not total social cost but the extra social cost from recycling an extra unit of waste. (In other words, it is the first derivative of a total social cost curve.) There are considerable advantages of working with marginal functions, as will be seen. MSC_R begins below the horizontal axis and rises above it. What this means is that it is initially socially profitable to recycle (revenues from the sale of materials exceeds the cost of recycling) but that after the point $R_{R\pi}$ it ceases to be socially profitable: costs are greater than benefits. This *a priori* expectation arises because extra recycling will involve collecting waste from more and more diffuse sources and perhaps processing lower and lower quality waste.

Now consider exactly the same analysis for landfill. In this case we assume that the marginal social cost of landfill, MSC_L, is always positive. This need not be so. It is possible that the sale of recovered energy from methane capture could exceed the actual costs of landfill, in which case MSC_L would begin below the axis in the same way as the recycling curve. Analysis of actual costs suggests this is unlikely—see Section 4. Note also that MSC_L includes any environmental impacts from landfill, so that it can be thought of as being made up of the actual costs of disposal and the environmental costs.

How much recycling should there be? In terms of Figure 1 the economically 'optimal' amount of recycling is given by the intersection of the two marginal curves at point B, giving an optimal recycling level of R*. The reason for this is that, at this point, the *total* social costs of disposal (inclusive of recycling) is minimized for society as a whole. Since areas under marginal curves are totals, this minimum social cost is represented by the area $R_{R\pi}BR^* + R^*BC\bar{W} - ODR_{S\pi}$ in Figure 1, *i.e.* the cost of recycling plus the cost of landfill minus any social profits from recycling. The equation of the two marginal curves thus has an important property: it minimizes the total social cost of disposal. For economists, this is the objective of IWM, and note that it includes in the concept of cost any environmental losses or gains (gains are simply negative losses). Few discussions of IWM in practice offer this as the objective, yet it is fundamental if there is to be any rationality in IWM. Failure to honour cost minimization amounts to throwing money needlessly at a problem; money that could be used to secure some other social benefit.

Figure 1 can be generalized for more than one final disposal option and for energy recovery in landfill and incineration.[2] What is important is that it shows the optimal level of recycling to be quite different from the level that would be determined by considering the revenues and costs of recycling. Suppose, for example, that there are no external costs from recycling. Then the workings of the market would produce an amount of recycling $R_{S\pi}$, *i.e.* the amount that would set marginal profits equal to zero, or, the same thing, the amount which maximizes total profits for the recycling industry. This level of recycling is too low from a social standpoint. Equally, what is socially desirable is not the maximum amount of recycling. That clearly imposes costs on society higher than those achieved by the optimal amount of recycling.

Figure 1 also suggests a way in which the optimal amount of recycling can be secured. It is clearly not going to emerge from the workings of the free market. But if recyclers were given a payment equal to BR*, it can be seen that the market

---

[2] Detailed analysis may be found in D. W. Pearce, I. Brisson, and R. K. Turner, 'The Economics of Waste Management', forthcoming.

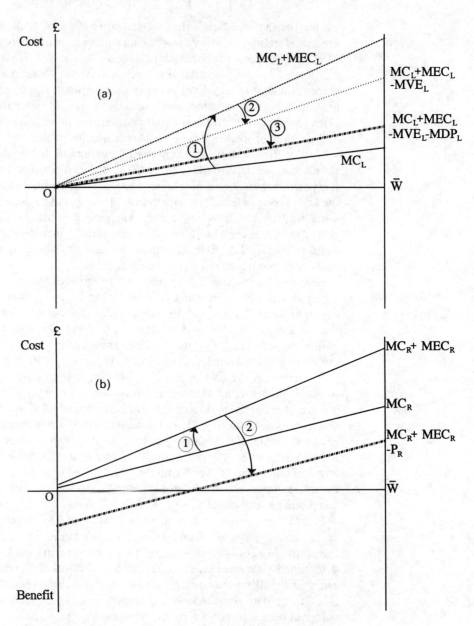

Figure 2 (a) Marginal social cost of landfill (MC_L is the marginal actual cost of landfill, MEC_L is the marginal cost of 'externalities', MVE_L is the marginal value of energy recovered, and MDP_L is the marginal cost of the pollution from the displaced energy source); (b) Marginal social cost of recycling (P_R is the price received for recycled waste)

would take the industry to a recycling level R*, the optimum. Is there a rationale for paying recyclers BR*? There is, because by recycling 1 tonne of waste, the waste disposal authority saves the disposal cost MSC_L. If disposal authorities therefore pay recyclers the cost of disposal which they save whenever more recycling is carried out, an optimal level of recycling will emerge. This is the basic idea behind the 'recycling credit' in the United Kingdom, discussed in Section 5.

Figure 2 shows some of the real world complexities that can be introduced into the analytical structure of Figure 1. We still retain the assumption of only one disposal option, landfill, because the introduction of other options makes the analysis more involved, although the analysis is not altered. The additional factors included in Figure 2 are: the decomposition of landfill costs into financial costs of disposal, environmental costs of disposal, and the value of energy recovered from methane capture; and the decomposition of recycling costs into normal costs and environmental costs. (Note the danger of double counting: the external costs of landfill can also be thought of as a benefit to recycling. Either approach is correct, but the item cannot be counted twice.) In Figure 2(a) we show the decomposition of MSC_L. The actual costs of landfill, *i.e.* the manpower and capital equipment and land costs are shown as MC_L. To these must be added the costs of any externalities, MEC_L, such as leachate, carbon dioxide and methane emissions, odour, noise, congestion caused by trucks, and so on. But any energy recovered from methane counts as a negative cost, *i.e.* it lowers the cost curve by the amount MVE_L. Finally, and of some importance, if energy is recovered from methane capture, it may displace other energy sources in the economy. For example, if it is converted to electricity it will displace the least efficient amount of electricity in the grid system, since that is how electricity supply systems operate—with the most efficient sources being used first and the least efficient being used last. Note that the value of the energy displaced is not relevant here, but the *pollution* from the displaced energy source is relevant. It is a benefit to the methane recovery system. This is equivalent to a further shift downwards in the cost curve for landfill and is shown in Figure 2(a) as a reduction by MDP_L.

This displacement effect also applies to other disposal options, and especially to incineration where the efficiency of energy recovery can be very high. We illustrate the importance of this effect in Section 4.

In Figure 2(b) the same analysis is shown for recycling. The marginal cost of recycling is always positive, and any externalities increase the cost. But the prices (P_R) received for materials sold will offset some of the cost, initially more than offsetting the cost but later failing to prevent costs from being positive. The heavy broken-lined curves in Figure 2(a) and (b) thus produce the cost curves in Figure 1.

Note that Figure 2 also permits some further investigation of the kinds of policies that might be introduced to secure the optimal amount of recycling, and hence the optimal amount of landfill. The 'recycling credit' discussed above could be extended to include any externalities from landfill. Since this makes the credit larger, more recycling will be undertaken. Alternatively, a tax could be levied on landfill equal to the marginal value of the externality imposed by landfill—the so-called 'landfill levy'. Finally, if energy recovery produces external benefits in excess of external costs, it may be legitimate to 'subsidize' an energy recovery system by the value of the net benefit it yields. 'Subsidy' is strictly not the correct word here since the payment to an energy recovery source would be a wholly legitimate payment for an external benefit, whereas subsidies are not payments for a benefit as such.

Figures 1 and 2 are essentially simple in concept. Yet they capture the essentials of the economic approach to waste management since they can be used to analyse the optimal level of recycling and the optimal level of final disposal. Moreover, they illustrate the potential role for various kinds of 'economic instruments' such as landfill taxes and recycling credits.

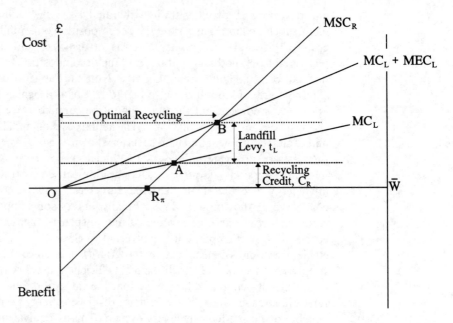

Figure 3 Optimal recycling and the rôle of recycling credits and landfill levies

Figure 3 shows more clearly how the credits and taxes apply to the IWM context. Figure 3 shows the marginal costs and marginal external costs of landfill (MC_L and MEC_L) and the marginal social cost of recycling. Assume, for convenience, that landfill is without energy recovery. Then, the optimal amount of recycling is at B. To secure this, a recycling credit of C_R can be given together with a landfill tax of t_L. Alternatively, t_L could be an additional component of the recycling credit.

4 Measuring Waste Disposal Externalities

The previous discussion sets out the basic theory of IWM. In this section we report estimates of the size of the various externalities associated with landfill land incineration in the UK. Estimates for recycling are not currently available.

The wide variety of types and scales of waste disposal facilities in the UK makes it impossible to define and evaluate a *typical* facility. To overcome this problem, an analytical framework has been developed in which a representative range of sites and conditions are defined and evaluated.[3] This range is represented by the following six scenarios:

L_1 existing *urban* landfill with *no* energy recovery
L_2 new *urban* landfill *with* energy recovery
L_3 existing *rural* landfill with *no* energy recovery
L_4 new *rural* landfill *with* energy recovery
I_1 new *urban* incinerator with energy recovery
I_2 new *regional* incinerator with energy recovery

---

[3] Full details may be found in CSERGE, Warren Spring Laboratory and EFTEC, 'The Externalities of Landfill and Incineration', HMSO, London, 1993.

Identification of Externalities

The main environmental impacts from landfill and incineration are associated with emissions to land, air, and water. Air emissions are particularly important as they include emissions from the transportation of waste as well as those from the disposal facilities themselves. In addition, air emissions from conventional energy sources can be 'saved' by the energy recovered from waste facilities. Resource use is also of concern both with regard to resource depletion and to energy use and recovery. Social impacts can include disamenities, such as noise, odours, visual impacts, and health impacts. The latter have been partially included in this study by undertaking an assessment of the accidents caused by the transportation of waste.

In an analysis of this type there are three stages of data analysis; firstly that of identifying the externalities, secondly the quantification of the impacts, and thirdly that of attributing a monetary value to the impacts. It has proved very difficult in this assessment to include a quantification and valuation of all the identified externalities. However, those considered to be the most important have been incorporated into the model, including gaseous emissions (from the disposal facilities, transport, and those associated with pollution displacement), leachate (emissions to land and water), and safety.

Both the landfilling and incineration of waste result in gaseous emissions of various forms. One category is greenhouse gases, contributing to global warming. Landfilled waste emits both carbon dioxide (CO_2) and methane (CH_4), two significant greenhouse gases. Where the CH_4 is captured and burnt for energy production it oxidizes to CO_2. Incineration of waste results in CO_2 emissions, but also 'conventional' air pollution in the form of sulfur dioxide (SO_2), nitrogen oxides (NO_x), and particulates (TSP). Further air pollution in the form of CO_2, NO_x, and TSP is caused by the transportation of waste from the catchment areas to the disposal facilities.

Where energy recovery is carried out, the electricity generated (it is assumed that energy is recovered in the form of electricity rather than as heat or combined heat and power) displaces that generated by other fuels. In this study the displaced electricity is taken to be that generated at the most inefficient power station, which currently in the UK is assumed to be an old coal-fired power station. It has also been assumed that the externalities associated with coal mining and power generation will be reduced in proportion to the energy saved. Thus energy recovery *saves* emissions of CO_2, NO_x, SO_2, TSP, and CH_4.

In addition to pollution of air, the landfilling of waste also results in pollution of land and water in the form of leachate, which is the product of the infiltration of precipitation into landfill, coupled with the biochemical and physical breakdown of wastes. This leachate can percolate down through the landfill and there is a danger that it might enter aquifers and surface waters and contaminate drinking water supplies.

Externalities other than pollution of air, land, and water include safety and congestion impacts of the transportation of waste. The health impacts from traffic emissions are of increasing concern, especially with the generally reported rise in asthma in urban areas. It has not been possible to include these impacts in the study, but it is an area which will no doubt be a topic of future research. An

Table 2 Summary of physical impacts from actual landfilling (mean values or best estimates only)

	Landfill scenarios:[a] emission impact/(tonne waste)$^{-1}$			
	L_1	L_2	L_3	L_4
Disamenity	not available on tonne^{-1} basis			
Leachate	not available on tonne^{-1} basis			
Global air/tonne				
CO_2 as C	0.024	0.035	0.024	0.035
CH_4	0.033	0.019	0.033	0.019
Transport:				
Pollutants/g(tonne waste)$^{-1}$(journey)$^{-1}$:				
CO_2	2256	2256	17 333	17 333
NO_x	30.3	30.3	334	334
TSP	4.3	4.3	8.8	8.8
Casualties/(tonne waste)$^{-1}$(journey)$^{-1}$				
Killed	0.15×10^{-6}	0.15×10^{-6}	0.38×10^{-6}	0.38×10^{-6}
Killed or seriously injured	0.77×10^{-6}	0.77×10^{-6}	1.88×10^{-6}	1.88×10^{-6}
Slightly injured	2.51×10^{-6}	2.51×10^{-6}	6.13×10^{-6}	6.13×10^{-6}
Displaced energy/kWh (tonne waste)$^{-1}$:		79		79
Recovered energy:		nil		nil
Displaced pollution/kg:				
CO_2 as C		23.4		23.4
NO_x		0.4		0.4
SO_2		1.1		1.1
TSP		0.01		0.01
CH_4		0.3		0.3

[a] See Section 4 of text for definitions of L_1–L_4.

area that has been possible to include is the incidence and cost of casualties associated with waste transport based on data published annually by the Department of Transport. Congestion impacts were not included in the original study reported here[3] but we include some illustrative examples further on.

Finally, landfill sites as well as incineration plants have a number of local impacts. Areas of concern raised at public inquiries on the siting of landfill sites included increased volumes of traffic, odour, health risks, wind blown litter, loss of visual amenity, and noise. With respect to incineration plants, the concern is particulary related to the health effects of emissions. These factors can affect the value of local properties. The degree of value reduction and the spatial extent of its influence is open to debate. Several US studies have sought to identify this external cost. Clearly the magnitude of this disamenity will depend in the first instance on whether we are dealing with an existing site or a proposed new site. The property blight effect is likely to be more marked in the latter than the former situation.

Tables 2 and 3 summarize the physical impacts from current landfill and incineration.

'Fixed' versus 'Variable' Externalities. Most of the externalities listed above can be related to the throughput of waste and thus be described as 'variable' externalities. Some externalities, however, such as disamenity, cannot be related

Table 3 Summary of physical impacts from actual incineration (mean values or best estimates only)

	Incineration type[a]/(tonne waste)$^{-1}$	
	I_1	I_2
Disamenity	not available on tonne^{-1} basis	
Global air/tonne C	0.19	0.19
Conventional air/tonne:		
TSP	96×10^{-6}	96×10^{-6}
SO_2	680×10^{-6}	680×10^{-6}
NO_x	1100×10^{-6}	1100×10^{-6}
Air toxics	not estimated	not estimated
Transport:		
Pollutants/g(tonne waste)$^{-1}$(journey)$^{-1}$:		
CO_2	7509	14 741
NO_x	124	275
TSP	8.7	9.8
Casualties/(tonne waste)$^{-1}$(journey)$^{-1}$:		
Killed	0.14×10^{-6}	0.22×10^{-6}
Killed or seriously injured	0.69×10^{-6}	1.11×10^{-6}
Slightly injured	2.27×10^{-6}	3.63×10^{-6}
Displaced energy:		
Recovered energy/kWh (tonne waste)$^{-1}$:	664	664
Displaced pollution/kg:		
CO_2 as C	195	195
NO_x	3.5	3.5
SO_2	9.3	9.3
TSP	0.1	0.1
CH_4	2.7	2.7

[a]See Section 4 of text for definitions of I_1 and I_2.

to tonnage in this fashion as it is the very existence of the facility that causes disamenities. While it is probably of some significance whether the annual throughput is 5 tonnes, 100 tonnes, or 5 million tonnes, it would be very difficult to measure disamenity effects in terms of £ (tonne of waste)$^{-1}$. Hence these externalities can be described as 'fixed' and a more appropriate way of attributing costs would be per site or per household affected.

Thus, in summary, we have the externality per site as:

$$E = F + (VQ) \qquad (1)$$

where E is the total externality, F is the fixed component, V is the variable component, and Q is the throughput.

In summary we have:

WASTE DISPOSAL EXTERNALITY = SITE DISAMENITY COST
[£ site^{-1} or household^{-1}]

and

GLOBAL POLLUTION COSTS + CONVENTIONAL AIR POLLUTION

COSTS + AIR TOXICS + LEACHATE COSTS

+ TRANSPORT RELATED COSTS − DISPLACED POLLUTION BENEFITS

= NET VARIABLE EXTERNAL COSTS [£ (tonne of waste)$^{-1}$]

Economic Valuation

Various methods can be used to attach monetary values to external costs and benefits. These include dose–response relationships, contingent valuation, and hedonic pricing.[4] A dose–response function associates a given level of pollution with a change in output which is then valued at market, revealed/inferred, or shadow prices. For example, dose–response functions have been used in other studies to look at the effect of pollution on health, physical depreciation of material assets such as metal and buildings, aquatic eco-systems, vegetation, and soil erosion. The dose–response functions are then multiplied by the unit value to give a monetary damage function. This approach is not suitable when valuing an externality such as disamenity because there is no direct relationship between the 'dose' of landfill/incineration taking place and the resulting disamenity experienced by neighbours to the sites.

More suitable valuation methods would be contingent valuation or hedonic pricing. In the former, people would be asked how much they would be willing to pay not to have an incinerator or landfill in their neighbourhood, or alternatively, what is the least they would be willing to accept (as compensation) for putting up with a waste disposal facility near their home. A hedonic property price study would attempt to measure the effect of the proximity of a waste disposal site on house prices. In the USA a number of studies using hedonic property pricing have been carried out for landfills. These studies, covering both municipal and hazardous waste sites, generally found that house prices rose between 5% and 10% (mile distance)$^{-1}$ from a site for up to four miles distance from the sites.[5–10] Unfortunately, the study reported here[3] was precluded from such analysis due to budgetary and time constraints.

The economic parameter values given in Table 4 are obtained from a range of

[4] For a comprehensive survey see A. M. Freeman III, 'The Measurement of Environmental and Resource Values—Theory and Methods', Resources For the Future, Washington DC, 1993.

[5] A. C. Nelson, J. Genereux, and M. Genereux, 'Price Effects of Landfills on House Values', *Land Econ.*, 1992, **68** (4).

[6] J. Havlicek, R. Richardson, and L. Davies, 'Measuring the Impacts of Solid Waste Disposal Site Location on Property Values', *Am. J. Agric. Econ.*, 1971, 53.

[7] J. Havlicek, 'Impacts of Solid Waste Disposal Sites on Property Values', Environmental Policy: Solid Waste, Cambridge, MA, 1985, 4.

[8] K. J. Adler, Z. L. Cook, A. R. Ferguson, M. J. Vickers, R. C. Anderson, and R. C. Dower, 'The Benefits of Regulating Hazardous Disposal: Land Values as an Estimator', US Environmental Protection Agency, Washington DC, 1982.

[9] H. B. Gamble, R. H. Downing, J. Shortle, and D. J. Epp, 'Effects of Solid Waste Disposal Sites on Community Development and Residential Property Values', 1982, Institute for Research on Land and Water Resources, Pennsylvania State University.

[10] R. Mendelsohn, D. Hellerstein, M. Huguenin, R. Unsworth and R. Brazee, 'Measuring Hazardous Waste Damages with Panel Models', *J. Environ. Manage.*, 1992, 22.

Table 4 Economic parameter values used in the estimates of landfill and incineration externalities

	£ tonne^{-1}
Global pollutants	
CO_2 as C	4.1–31.0
CH_4	31.9–138.5
Conventional pollutants[a]	
SO_2	425
NO_x	327
TSP	14 221
Conventional pollutants[b]	
SO_2	245
NO_x	83
TSP	14 221
Leachate	
Existing landfills	0–0.9
New landfills	0
Casualties	
£ (Mortality)$^{-1}$	0.715–2.0 million
£ (Serious injury)$^{-1}$	74 780
£ (Minor injury)$^{-1}$	6 080

[a]Transboundary damage; [b]Damage to the UK only.

Table 5 Summary of externality values for landfill [£(tonne waste)$^{-1}$ other than disamenity]

Landfill scenarios*	L_1	L_2	L_3	L_4
(a) Global pollution				
CO_2 as C	0.32	0.46	0.32	0.46
CH_4	2.36	1.36	2.36	1.36
(b) Air pollution	not applicable	not applicable	not applicable	not applicable
(c) Transport impacts				
Pollution**	0.09	0.09	0.38	0.38
Pollution†	0.10	0.10	0.46	0.46
Accidents	0.27	0.27	0.60	0.60
(d) Leachate	0.45	0	0.45	0
(e) Pollution displacement**	0	0.81	0	0.81
(f) Pollution displacement†	0	1.12	0	1.12
Total (a + b + c + d − e)**				
Mean	3.50	1.38	4.11	1.99
Total (a + b + c + d − f)†				
Mean	3.50	1.08	4.19	1.76

*See Section 4 of text for definitions of L_1–L_4.
**Conventional air pollution including damage to the UK only.
†Conventional air pollution including transboundary damage.
These estimates omit any disamenity costs which may well be significant.

different studies which, given the scope of this paper, are not discussed here.[11] The physical parameter values of Tables 2 and 3 are multiplied by the economic parameter values of Table 4 to estimate the externality values for landfill and incineration in Tables 5 and 6.

[11] For further details and discussion see [4] and D. W. Pearce, C. Bann, and S. Georgiou, 'The Social Costs of Fuel Cycles', HMSO, London, 1992.

Table 6 Summary of externality values for incineration [£(tonne waste)$^{-1}$ other than disamenity]

Incineration scenarios*	I_1	I_2
(a) Global pollution		
CO_2	2.55	2.55
CH_4	not applicable	not applicable
(b) Air pollution		
Conventional**	1.62	1.62
or		
Conventional†	2.01	2.01
Toxics	not estimated	not estimated
(c) Transport impacts		
Pollution**	0.23	0.36
Pollution†	0.26	0.42
Accidents	0.37	0.53
(d) Pollution displacement**	6.87	6.87
(e) Pollution displacement†	9.40	9.40
Total (a + b + c − d)**		
Mean	−2.10	−1.81
Total (a + b + c − e)†		
Mean	−4.21	−3.88

*See Section 4 of text for definitions of I_1 and I_2.
**Conventional air pollution including damage to the UK only.
†Conventional air pollution including transboundary damage.
These estimates omit any disamenity costs which may well be significant.

Tables 5 and 6 summarize the results of the study. The totals reflect the sum of all the externalities, excluding disamenity effects. They indicate that for landfill the lowest external costs, in the order of £1 (tonne of waste)$^{-1}$ in an European context,[12] slightly more in a UK context,[13] are associated with urban sites which have low transport impacts and energy recovery (the global pollution costs are less for sites with energy recovery and, in addition, there are pollution displacement *benefits*). The highest estimate for external landfill cost is for rural landfill sites (high transport impacts) with no energy recovery (high global pollution and no benefit from displaced pollution) and is around £4 (tonne of waste)$^{-1}$.

For incineration, Table 6 shows estimated overall external *benefits* as the benefits from displaced pollution resulting from energy recovery outweigh the external costs of global and conventional pollution and transport impacts. When only effects on the UK are taken into account, this net benefit is in the region of £2 (tonne of waste)$^{-1}$; slightly more for urban incinerators and slightly less for regional incinerators. When effects for the whole of Europe, including Eastern Europe and Scandinavia, are included the net benefits rise to around £4 (tonne of waste)$^{-1}$; again slightly more for urban incinerators while slightly less for regional incinerators.

[12] Including transboundary effects of 'conventional' pollutants (NO_x and SO_2) from transport and saved transboundary effects from displacement of other energy sources for all of Europe.
[13] Effects of 'conventional' air pollutants from transport and saved in the context of energy displacement are considered for the UK only.

Thus landfilling causes net external costs while incineration gives rise to net external benefits indicating that *based on externalities alone* a diversion of waste from landfill to incineration would be beneficial.[14] However, it must be kept in mind that (a) these figures are only single points on the cost function for each waste disposal method—as incineration increases, the external benefits might turn into external costs, and (b) external costs are only part of the overall costs of disposal. A further caveat is that no estimation of the disamenity effects of landfill and incineration has been carried out. In some countries there is very strong resistance to the siting of incineration facilities as there are strong fears of the perceived health effects. If this is not matched by an equal resistance to landfills, adding this component to the present results may alter the overall result to the extent that the external costs from incineration exceed those from landfill.

Congestion Costs. One omission in the estimation of impacts of the transportation of waste to the disposal facilities in the CSERGE *et al.* study was the congestion disbenefit. The other two impacts estimated, air pollution and accidents, were less for urban sites than for rural or regional sites because of the shorter distance between the catchment area and the urban site than between the catchment area and the rural and regional sites. Congestion, however, is generally a greater problem in urban than in rural areas. The question is, how important a problem it is, and whether the congestion costs associated with urban sites would outweigh the advantages of the shorter distance in terms of air pollution and accidents.

A recent study[15] estimates the marginal congestion costs for a wide range of roads and conditions, from urban central peak [36.37p (passenger car unit kilometre)$^{-1}$ (PCUkm)] to other rural roads [0.05p (PCUkm)$^{-1}$], with a weighted average of 3.40p (PCUkm)$^{-1}$. To make these data applicable to the transportation of waste, we would need the marginal congestion cost (HGV unit km)$^{-1}$ (HGVUkm). The numbers we have used to calculate air pollution and accident costs reflect different sizes of HGVs, but as a rough guide we use 2 PCUkm = 1 HGVUkm.[16]

Owing to the wide range of values, it is important to ensure that the waste transportation is placed in the right road category. If we assume that the transport of waste to urban landfills and incinerators takes place on central urban roads in the off-peak, we can calculate a congestion cost of £3.65 (tonne of waste)$^{-1}$ going to landfill and £6 (tonne of waste)$^{-1}$ going to incineration. If we, alternatively, assume that the transport takes place on roads of the 'other urban' category (presumably quiet residential streets), these figures fall to £0.01 and £0.02 (tonne of waste)$^{-1}$, respectively.

These figures illustrate that congestion costs can, under certain circumstances,

---

[14] However, in order to make such a comparison, *total* costs including external as well as financial costs must be compared. As current financial costs of landfill range from $7.5 to $22.5 compared to $20–30 for incineration, the chances are that even when including the external costs of landfill and the external benefits of incineration, landfill will still in most cases come out as the less costly option.

[15] D. M. Newbery, 'Pricing and Congestion: Economic Principles Relevant to Pricing Roads', *Oxford Rev. Econ. Policy*, 1990, **6** (2), pp. 22–38.

[16] As suggested in D. M. Newbery, 'Road User Charges in Britain', *Econ. J.*, 1988, **98**, 161–76.

Figure 4 The inefficiency of recycling targets

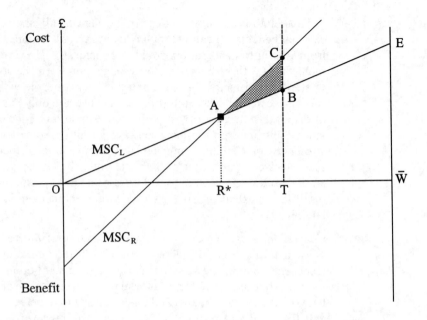

add considerably to total external costs, while under other circumstances they would be insignificant. Therefore an accurate estimation of the congestion costs would require careful modelling of the respective transport routes of the waste. Most (municipal) waste is *collected* on quiet residential roads ('other urban', 'small town', and 'other rural') but, when taken for disposal, will probably be transported along busier roads to the landfill, incinerator or transfer station.

5 Economic Instruments for Integrated Waste Management

Once estimates for the externalities are made, the final issue is how best to implement policy. The most popular mechanism is through the setting of targets, *e.g.* X% of material Y should be recycled by year t; 90% of all waste should be recycled or re-used, and so on. Unfortunately, while targets are appealing to politicians, they can often be very inefficient in that they waste resources. To see this, Figure 4 repeats Figure 1 but showing a target T for the level of recycling. The optimal level of recycling is R* as before, but T sets a target in excess of this. The cost of going beyond the optimum level of recycling to recycle R*T is area R*TCA, but R*T would be more cheaply dealt with by landfill, even allowing for all the externalities. Since the landfill cost would have been R*TBA, the excess cost (known as the 'deadweight' cost) is the shaded area ABC.

Targets can easily be inefficient. Finding the optimal level of recycling is to be preferred. However, even when targets are unavoidable, charges, taxes, or 'tradeable recycling obligations' are usually cheaper ways of achieving the target than simply requiring everyone to reach the target. In general, these approaches, known as 'economic instruments', are less costly for industry and consumers whilst still protecting the environment. We briefly survey the main instruments.

Landfill Levies

A landfill levy, or tax, has been shown to be one mechanism for securing optimal recycling. In the UK the Government is considering introducing a landfill levy. Although no final decision has yet been made public regarding the form or size of a landfill levy, some indication can be found in reports commissioned by the Department of Environment.[17] A uniform levy for all types of waste and landfills is recommended on the grounds of ease of administration. With respect to the size of a levy, the Department of the Environment suggests, on the basis of the estimates of externalities from landfill and incineration, that a levy on landfill might be set as the difference between the estimated external cost of landfill and the estimated external benefit of incineration, i.e. within a range of £5–8 (tonne of waste)$^{-1}$ if transboundary effects are taken into account, or £3–6 (tonne of waste)$^{-1}$ if only UK effects are considered. The rationale for placing the entire difference in external costs and benefits of landfill and incineration, respectively, as a levy on landfill, rather than placing a levy equal to the external costs of landfill on landfill and providing a subsidy equal to the external benefit of incineration to incineration, is that it is the relative prices that matter. The introduction of a, say, £8 levy (tonne of waste)$^{-1}$ going to landfill or a £4 levy on landfill and a £4 subsidy to incineration would, if landfill and incineration existed in isolation, lead to the same amount of waste being diverted from landfill to incineration. So in that situation the presumption made by the Department of the Environment would be correct.

However, if other means of waste treatment exist, such as recycling, the former option would be inefficient because recycling relative to incineration would be cheaper than if a subsidy of £4 tonne^{-1}, reflecting the external benefit, were paid to incineration. Thus the former option would lead to more waste being diverted from landfill to recycling than warranted, at the expense of incineration.

From a revenue point of view, it is also highly relevant whether the externalities are translated into a landfill levy of £5–8 tonne^{-1} (£3–6 tonne^{-1} for UK effects only) or a landfill levy of £1–4 and an incineration subsidy of £2–4 tonne^{-1}. In the current situation where 102 million tonnes of controlled waste are landfilled and 4 million tonnes are incinerated, the former option would raise revenues of £510–816 million year^{-1} (£306–612 million year^{-1} if only UK effects are internalized), whereas the latter option would raise revenues of only £112–428 million year^{-1} (£143–418 million for UK effects only), while subsidies of £16–34 million year^{-1} (£7–8 million for UK effects only) would be required. In fact, if the effect of the internalization of the external costs and benefits of landfill and incineration were to be a substantial diversion of waste from landfill to incineration, it could even be envisaged that the subsidies required for incineration would exceed the revenues raised by the landfill levy. On this background, it is maybe understandable why the former option would be preferred from a Government point of view.

---

[17] See foreword to [4] and Coopers and Lybrand 'Landfill Costs and Prices: Correcting Possible Market Distortions', HMSO, London, 1993.

Kerbside Charges

In the case of household waste, the introduction of landfill levies will not do much to change household behaviour unless the levy is passed on to households. In the UK, as in most other industrialized countries, waste services do not carry a unit price; instead it is financed over the local tax bill or through property taxes, charging the households a flat rate which does not reflect the quantities of waste put out for disposal. Thus, for the household, the marginal cost of putting out an extra bag of waste is zero, and hence there is no economic incentive to reduce the amount of waste generated. A number of local communities (*e.g.* in the US, Germany, the Netherlands, and Denmark) have introduced waste service charging schemes, whereby the householder pays according to quantity of waste disposed of. This can either be in the form of pre-paid rubbish bags, a choice of size of bins and frequency of collection, or by weight. Evidence shows that considerable reductions in waste generation can be achieved. A kerbside charge can reflect just the financial costs of collection and disposal of waste (in Figure 3 this is represented by C_R) or, if more ambitious, can include the external cost as well. In Figure 3 this would be $C_R + t_L$.

Recycling Credit Scheme

In the 1990 Environmental Protection Act the UK Government introduced what are popularly known as recycling credits; an economic incentive-based approach to increasing recycling. According to the Act, collectors of household waste for recycling should receive payments based on the financial savings arising from reduced waste collection and disposal costs. Thus the payments reward the setting up of recycling collection schemes. The payments are made by the Waste Disposal Authorities and Waste Collection Authorities who are saving on not having to collect or dispose of waste that is recycled. As a result, the recycling credit is not a drain on the central government budget, as it is merely a transfer payment between different tiers of local government.

For a graphic illustration of recycling credits consider Figure 3. In a market where no recycling credit is paid and no landfill levy is charged, recycling will take place until the point where private profits turn into a loss. If we, for simplicity, assume that there are no external costs associated with recycling this will happen in Figure 3 at R_π. However, R_π is less than the socially optimal level of recycling, B. By making the *financial* cost saved through recycling available to recyclers through the recycling credit, C_R, the amount of waste recycled increases from O–R_π to O–A. This is obviously still below the socially optimal level of recycling, O–B, which also takes into account the *external* cost savings, but all the same a 'better' solution than leaving the market to its own devices.

Product Charges

Alternative to 'downstream' measures such as recycling credits, landfill levies, and kerbside charges, the waste problem can also be addressed 'upstream', for instance through a product charge to counter command-and-control policies such as recycling targets. This approach involves assigning collection and disposal costs, financial as well as external, to individual products and incorporating them into a product charge. The advantage of a product charge is that it signals to the consumers, not only the cost of production, but also the cost of disposal, thus allowing the total costs of different product to be compared.

On this basis the general pricing rule for a given product becomes:

$$P = MPC + MEC + MLUC \qquad (2)$$

where P is the price of the product; MPC is the marginal private cost of production, MEC is the marginal external cost (collection, disposal, and litter), and $MLUC$ is marginal landfill user costs.

This can be further expanded to:

$$P = MPC + MCC + MDC + MLC + MLUC \qquad (3)$$

where MCC is marginal collection costs, MDC is marginal disposal costs, and MLC is marginal litter costs. $MLUC$ will be related to MDC as follows:

$$MLUC = MDC_T(1 + r)^{-T} \qquad (4)$$

where T is the time at which some replacement disposal route has to be found to replace exhausted landfill sites, and r is the discount rate.

Equation (3) thus becomes generalized as:

$$P_t = MPC_t + MCC_t + MDC_t + MLC_t + MDC_T(1 + r)^{-T} \qquad (5)$$

where t is time. If MCC, MDC, MLC, and $MLUC$ are not already incorporated into the price of the product through regulation, the product charge, τ, that is needed is:

$$\tau = MCC_t + MDC_t + MLC_t + MDC_T(1 + r)^{-T} \qquad (6)$$

To encourage source reduction and recycling, two further elements can be added and thus make the product charge superior to arbitrary recycling targets:

$$\tau = (W_i/U_i)(1 - r_i)MCC_t + MDC_t + MLC_t + MDC_T(1 + r)^{-T} \qquad (7)$$

where W_i is the weight (or the volume) of the ith product; U_i is the unit that the ith product comes in, e.g. litres, boxes, bags, m^3, etc., so that W/U is weight unit^{-1} of the product; and r is the recycling rate of the ith product.

The first derivatives tell us:

$\partial \tau/\partial W > 0$ as the weight (or volume) of the product increases the product charge increases; as the weight goes down (light-weighting/down-sizing) the product charge comes down;

$\partial \tau/\partial r < 0$ as recycling goes up the product charge falls;

$\partial \tau/\partial C > 0$ where $C = MCC + MDC + MLC + MLUC$, i.e. as damage costs rise the product charge rises.

Thus the product charge achieves source reduction, recycling, and a lower environmental impact via changes in the mix of products available in the market without mandating arbitrary recycling targets. However, although a product charge has the advantage, as pointed out earlier, that it allows consumers to compare the total production and disposal costs of different products and thus offers them an incentive to adjust their behaviour accordingly, it has the disadvantage that high levels of information are essential. Not only is it necessary to assign the different cost element to each product which can become a very complex task indeed.

Deposit–Refund Schemes

Deposit–refund schemes (DRSs) are only relevant in a limited number of cases as it would hardly be feasible to operate a DRS for all waste. Also, economics tells us that DRSs should only be employed where the benefits outweigh the costs. This can either be the case in a market-generated system, *e.g.* beverage containers, where the costs of operating the system are less than the expected overall revenues to the producer. This is usually because V (the net reuse value of the scrap item) is positive, or because the refund, R, stimulates a significant increase in demand, sufficient to offset a negative value of V. Alternatively, schemes can be imposed by law in cases where the social benefit of retrieving a product outweighs the costs of operating the system. This is particularly relevant in the case of hazardous materials, which, if disposed together with ordinary mixed waste, can create unacceptable hazards. A DRS will help retrieve such materials for safe disposal.

More formally, DRSs should only be implemented where the net present value of the system is positive.[18]

$$NPV = \sum_{t=0}^{T} \frac{(B_{WSt} + B_{HSt} + B_{LSt} + B_{PSt} - C_{It} - C_{Ct})}{(1+r)^t} \qquad (8)$$

where NPV is net present value; B_{WS} is expected social benefit from a reduction in the amount of waste requiring collection and disposal; B_{HS} is expected social benefit from reducing the hazard of co-disposal; B_{LS} is expected social benefit from a reduction in the amount of litter, which in turn comprises $B_{LS} = bpc + bag$, of which bpc is reduced costs of litter pick-up and bag is the amenity gain from reduced litter; B_{PS} is the expected social benefit from a reduction in the quantity of inputs required for the production of the product subject of the scheme (*i.e.* reduced labour, materials, and energy costs) if the collected waste is recycled; C_I is expected social costs due to an increase in the quantity of inputs required in storage, handling, and distribution; C_C is expected social costs due to an increase in the time spent by households in returning the used product, *i.e.* consumers' inconvenience costs; and r is the discount rate.

Actual DRSs have generated both litter and disposal cost savings, although the exact magnitude of these social gains have varied from scheme to scheme and have proved difficult to value to everyone's satisfaction. Handling, storage, and transport costs also seem to have been significant.

Evidence from a number of DRSs for beverage containers has also indicated that return rates are not very sensitive to the size of the deposit. A much more important factor in this context has been the number, knowledge, and

---

[18] For further analysis and evaluation of deposit-refund system see I. Brisson, 'Packaging Waste and the Environment: Economics and Policy', *Resources, Conserv. Recycl.*, 1993, **8**, 183–292.

Figure 5 Tradeable recycling obligations (A and B indicate two different waste disposal authorities)

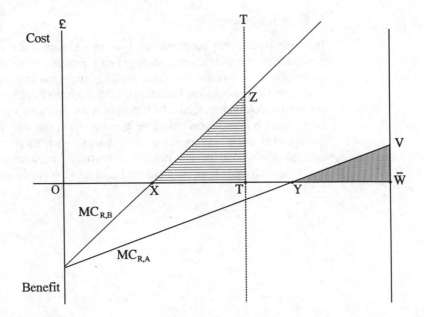

convenience of container return points. Inconvenience costs to consumers may well fall over time as individual adjust to the returnable system. Government/legislation could mandate the required number and type of returning points which, in turn, would boost return rates. The downside of this approach is that the greater the number of returning points the higher will be the overall system costs for handling, storage, and transport of returns.

Tradeable Recycling Obligations

While not so far tested in practice, a tradeable recycling obligation (TRO) can also be used to secure efficiency in waste management. Figure 5 illustrates the basic idea. Once again the same diagram is used but this time we assume two waste disposal authorities, A and B. Their cost curves are as shown. For convenience we omit all externalities and the disposal cost curve is assumed to be the same for each authority and is not shown. If A and B leave the market to determine how much recycling will be done they will settle at points X and Y, respectively (where the recycling industry maximizes profits). Now let the government establish a target T between X and Y. To meet the target authority B has to expand its level of recycling by XT, whereas A already achieves the target. Nonetheless, A has lower recycling costs than B. What B might do then is pay A to recycle XT instead of doing it itself. This can be shown to be profitable for both parties. If B recycled XT it would cost the shaded areas XTZ. But if A takes on this obligation on top of its own level of recycling, it will cost YW̄V which is less. It follows that as long as B pays A more than YW̄V, A will be better off. B will be better off as long as its payment to A is less than XTZ which is the cost it will otherwise have to bear. The two parties can therefore 'trade' an obligation and keep the costs of meeting the target to a minimum.

6 Conclusions

We have argued that economic analysis provides the most coherent and logical approach to integrated waste management. Popular ideas that 'rank' options in terms of source reduction, re-use, recycling, incineration, and landfill (usually in that order) have no logical foundation, although the ranking might turn out to be correct on detailed analysis. In the same vein, the idea that 'more recycling is better' has no foundation unless we are clear what the starting point is and what the relevant costs and benefits are. Finally, once targets have been agreed, preferably informed by cost–benefit comparisons, economic instruments such as credits and charges provide a cheaper and more effective way of securing those targets.

Subject Index

Activated sludge, 33
  plants (ASPs), 19
  process, 18
Activity of radioactive waste, 122
Adverse health effects, 87, 88
Aerobic treatment plants, 58
AIDS, 30
Aire and Calder project, 14, 15
Aluminium recycling, 98
Ammoniacal nitrogen (AmmN), 23, 25
Anaerobic,
  digesters, 54
  digestion, 24, 66, 73
    of sludges, 28
    plants, 20
  micro-organisms, 51
  microbial activity, 51
Appropriate,
  low-cost sewage treatment, 40
  technology, 72
Aquifer pollution by landfill leachate, 65
Asbestos wastes, 55
Atmosphere, 69, 70
Atmospheric emissions, 28
Atomic Energy Control Board, 117

Background radiation, 116, 127
BAT, 7, 9
Bathing Water,
  Directive, 29, 30
  quality, 29
BATNEEC, 2–8, 10, 71, 72, 74
Batteries, 99, 106
  Directive, 91
Battery recycling, 106
Best available,
  techniques, 72
    not entailing excessive cost (BATNEEC), 2, 71
  technology (BAT), 7
Best,
  practicable environmental option (BPEO), 2, 7, 10, 34, 66, 71
  practical means (BPM), 7
  practice, 8
  technical means (BTM), 7
Biochemical oxygen demand (BOD_5), 18
Biodegradable wastes, 69
Biogas production, 37
Biological,
  aerated filters (BAFs), 19
  aerated filtration (BAF), 33
  filters, 18, 19
  secondary treatment, 35
  treatment, 19
Biologically active waste, 53
Bioreactor, 69
BOD, 23, 40
Bottle banks, 100, 103
BPEO, 2, 10, 11, 72
BPM, 7
BPRO, 35
BTM, 7

Canadian Atomic Energy Control Board, 124
Canadian disposal concept, 118

Subject Index

Canadian Shield, 121, 122, 124, 125, 129, 130
CANDU,
  fuel, 119
    bundle, 116
  reactor, 116
Chemical,
  oxygen demand (COD), 23
  waste incinerator, 87
Chemically recycled plastics, 99
Chief Inspector's guidance notes, 9, 10, 14
Clean Air Act (1956), 50
Cleaner technology, 11, 12
Co-disposal, 44, 52–55, 66, 67
  landfill, 52
  practice in the UK, 54
  sites, 55
Coloured glass collection, 102
Combating of air pollution from industrial plants, 73
Completion criteria for leachates, 63
Compliance monitoring, 4
Composite landfill liner, 55
Composition of tyre rubber, 108
Compostable wastes, 106
Composting, 66, 73, 76
Congenital malformations, 87
Congestion costs, 145, 146
Containment landfill, 46, 48, 56
Contaminants,
  in recyclable materials, 97, 98
  in wastepaper, 98
Control of landfill gas emissions, 73
Cost,
  and benefit, 132, 134
  curve for landfill, 137
  of waste disposal, 65
Cost–benefit,
  analysis, 8
  comparisons, 152
  thinking, 133
Costs of disposal, 131
Council Directive on the Landfill of Waste, 43
Cullet, 99, 102, 103

Deposit–refund schemes, 150
Dilute and attenuate, 45
Disamenity effects, 145
Discount rate, 149
Disposal of nuclear fuel waste, 129
Dose–response relationships, 142
Drinking Water Directive, 63
Dry tomb, 57, 58, 69
Dust, 74, 77, 83, 84, 87

EA, 31, 32
EC,
  Dangerous Substances Directive, 24
  Directives, 21
  Environmental Policy, 21
Economic,
  analysis, 152
  instruments, 137, 146, 152
  theory, 131, 133
  valuation, 142
Economics of waste management, 131
Efficiency in waste management, 151
Emission budget, 82
Emissions to atmosphere, 69, 70, 77, 78, 82, 85, 87, 88
Entombment landfill, 48
Environmental,
  Agency, 21
  Assessment (EA), 31, 73
  assets, 132
  capacity, 10
  impact assessment, 31
  impacts, 131
  impacts from landfill and incineration, 139
  Protection Act (EPA 90) 1990, 1–3, 5, 6, 22, 27, 28, 35, 62, 71, 92, 111
  Protection Regulations 1991, 1
  Quality Objectives (EQO), 20
  Quality Standards (EQS), 20
  Statement (ES), 31
EPA 90, 5, 11, 65, 74
Epidemiological studies, 29, 88
ES, 31

Subject Index

European,
    Groundwater Directive, 46
    Union (EU), 73
Exposure pathways, 86
Externality values,
    for incineration, 144
    for landfill, 143

Faecal organisms, 30
Fail-safe landfill, 48, 49, 58
Ferrous,
    metal recycling, 104
    scrap, 104
Fifth Environment Action
    Programme, 43, 44, 67, 74
Flexible membrane liners, 56
Flue gas desulfurization (FGD), 7
Forestry, 27
Fugitive,
    emissions, 75, 76
        from landfill, 75
    gases, 60

Gasification, 111
Generally acceptable technology, 7
Geological disposal of nuclear fuel
    waste, 118, 119
Geomembranes, 56, 57
Geotextiles, 59
Glass, 99, 102
Global,
    emissions of trace elements, 79
    inventory of trace metals, 79
    warming, 60
Gram-negative bacteria, 76
Granulation of rubber, 109
Greenhouse gases, 61, 80, 81, 139
Groundwater,
    Directive, 46
    flow system, 128
    movement, 121, 123
    vulnerability, maps, 56

Hazardous waste incineration, 74, 79, 85, 86
Health,
    effects, 85, 88

    hazards, 131
    impacts from traffic emissions, 139
    statistics, 87
Heavy metals, 33, 99
Hedonic pricing, 141
Her Majesty's Inspectorate of
    Pollution (HMIP), 1, 3, 20, 71, 111
HIV, 30
HMIP, 1–3, 5–10, 13, 14, 20–22, 34, 71, 111
Household waste, 94, 95

Improvement programme, 4
Incineration, 28, 61, 73, 80, 112, 145
    externalities, 143
    of tyres, 111
    scenarios, 144
Industrial,
    effluent control, 33
    effluents, 17, 33
Integrated,
    gasification and combined cycle (IGCC), 13
Integrated Pollution Control (IPC), 1, 8, 34, 71
    waste management, 133, 146, 152
Intermodal Surface Transportation
    Efficiency Act 1991, 110
International,
    Atomic Energy Agency, 115
    Commission on Radiological
        Protection, 117
    toxic equivalents, 80
IPC, 1, 2, 6, 8–10, 14, 71–73

Kerbside,
    charges, 148
    collection, 100, 101
    schemes, 103

Land areas required for sewage treatment, 37
Landfill, 27, 43, 73, 75, 112, 135, 145
    costs, 137
    design, 55, 59
    Directive, 44, 64

Subject Index

(Landfill *cont.*)
  emissions of methane, 60
  gas, 52, 59, 60–62, 77, 78, 80, 81
    as a source of energy, 61
    composition, 53
    control, 59
    emissions, 74
    monitoring, 61, 62
    utilization, 60
  leachate, 46, 52, 58
    control, 59
  levy, 92, 137, 138, 147, 148
  licensing controls, 64
  liners, 56, 57
  management, 60
  practice, 45
  site closure and aftercare, 62
  with energy recovery, 61
Landfilling, 80
  tyres, 112
Landraising, 49
Leachate, 48
  drainage, 58, 59
  management, 57, 58
  recycles, 58, 59
  from domestic wastes, 54
Leptospirosis, 30
Levy of landfill, 147
List I Substances (Black List), 7, 47
List II Substances (Grey List), 7, 47
Locally unacceptable land uses (LULUs), 82
Love Canal, 86

Marginal,
  cost of recycling, 137
  social cost of landfill, 134–136
  social cost of recycling, 134, 136
Materials reclamation facilities, 101
Membrane,
  bioreactor systems, 39, 40
  filtration, 39
  separation, 39, 40
Methanogenic,
  bacteria, 51
  refuse, 55
  waste, 54

Microbial activity, 54, 76
Microbiological standards, 29, 30
Mitigation of impacts, 88
Monitoring, 4
Mono-disposal, 55
Municipal,
  solid waste, 76
  waste, 43
    arisings and disposal world-wide, 132
    incineration, 74

National,
  household waste analysis project, 94
National Rivers Authority (NRA), 3, 20, 46
Net external,
  benefits, 145
  costs, 145
Net social benefit, 8
Noise, 83, 84
  nuisance, 84
Non-fossil fuel obligation, 61, 112
Non-market price, 132
Non-prescribed processes, 72, 73
'Not in my back yard' (NIMBY), 82
NRA, 3, 20–23, 34, 46, 56
Nuclear fuel waste, 115, 117, 118, 120, 128, 130
  container performance, 122
  disposal facility, 120
  hypothetical disposal system, 125
  hypothetical repository, 125
  management program, 130
  repository, 121
    and fracture zone, 126
Nuisance complaints, 83
Nutrient-stripping tertiary treatment, 23

Odorous emissions, 83
Odour, 76–78, 82–84
  nuisance, 83, 84
  units, 76
Oil,
  laundering, 107
  wastes, 107

Subject Index

On-line monitoring, 39
Optimal recycling level, 135, 136, 146
Organic micropollutants, 76
Ozone-depleting gases, 61

Packaging,
  and Packaging Waste Directive, 91
  waste, 92
Paper recycling, 103
Pathogenic bacteria, 76
Physical impacts from,
  incineration, 141
  landfilling, 140
Physicochemical treatment, 70, 73
  plant, 83, 84
Plastics recycling, 105
Plutonic rock, 118, 119, 124, 129, 130
Plutonium, 115, 123, 127
Prescribed processes, 71
  and substances regulations 1991, 111
Priority waste streams, 92
Product charges, 148, 149
Public health, 28
  effects, 86
Pyrolysis, 111

Radiation, 116
  dose, 117
Radioactive,
  contaminant movement, 123, 124
  contaminant release, 122
  contaminants, 128
  decay, 117
  elements, 116, 117, 122, 123
  waste, 119
Radiological risk, 124, 127
Recyclable,
  components of household waste, 95
  material in household waste, 101
  waste, 100
Recycled plastics, 99
  rubber, 110

Recycling, 91–93, 96, 97, 102, 134, 135
  costs, 137
  credit, 136–138, 148
    scheme, 148
  of packaging wastes, 91
  targets, 146
  waste materials, 91
  waste-paper, 103
Regulatory framework in the UK, 71
Remediation, 62
Resource,
  conservation, 95, 96
  Conservation and Recovery Act, 57
Retreading tyres, 109
Royal Commission on,
  Environmental Pollution (RCEP), 1, 10, 112
  Sewage Disposal, 17
Rubber crumb, 109, 110
  reclaim, 109

Scrap,
  tyre recycling, 107
  tyres, 107, 109, 111–113
  use, 97
Sea disposal of sludge, 20, 27
Seveso, 86
Sewage, 17
  connections, 22
  discharges to watercourses, 34
  sludge, 24, 25, 34,
  treatment, 19, 22, 28, 31, 32, 37, 39
  undertaker, 34
Shadow prices, 132
Sludge, 24
  disposal, 25
  quality, 26
  treatment, 20
  (Use in Agriculture) Regulations 1989, 25
Socially optimal level of recycling, 148
Soil conditioners/fertilizers, 37
SS, 23

157

Subject Index

Stack emissions, 75, 77
Standard Waste Water Treatment
    Directive, 23
Substances,
    List I, 7, 47
    List II, 7, 47
Suspended solids (SS), 18
    limit, 23
Sustainable,
    development, 48, 55, 58
    landfill, 48, 49, 58

Technical guidance notes, 9
Textile fibre from tyres, 110
Toxic release inventory, 79
Tradeable recycling obligations,
    151
Tyre incinerators, 111
Tyres, 108

UK waste arisings, 93, 94
Uranium, 115, 116, 128
    dioxide, 122, 127
Urban Waste Water Treatment
    Directive (UWWTD), 21
UWWTD, 21–24, 31, 35

Volatile Organic Compounds
    (VOCs), 12

Waste
    arisings for the UK, 93
    arisings world-wide, 131
    characteristics, 50
    composition, 50
    degradation, 50, 51
    disposal externalities, 138
    incinerators, 79
    management,
        option, 52, 67, 82
        papers, 73
        practices, 94
        processes, 69
        site, 31
    minimization, 1, 14, 15
    oils, 107
    treatment processes, 72
    paper, 98, 103
Waste-to-energy,
    incineration, 66
    plants, 81
Wastewater,
    discharges, 29
    treatment, 29, 32
Water,
    Industry Act 1991, 21, 33
    Resources Act 1991, 21–23, 46
Weil's disease, 30
Willing to pay (WTP), 132